ISAMBARD KINGDOM BRUNEL

Colin Maggs MBE is one of the country's foremost transport and engineering historians and has written over one hundred books, including *A History of the Great Western Railway* ('Comprehensive and full of charming anecdotes', Christian Wolmar; 'Thoroughly researched and well written ... a very readable account', *BBC Who Do You Think You Are Magazine*). He lives in Bath.

ISAMBARD KINGDOM BRUNEL

THE LIFE OF AN ENGINEERING GENIUS

COLIN MAGGS

AMBERLEY

Author's Note

Other biographies of Isambard Kingdom Brunel have divided his work in categories such as the Thames Tunnel, docks, railways, ships and bridges. The present book has generally taken a chronological stance to show how his various projects were intermingled.

This edition published 2017

Amberley Publishing
The Hill, Stroud
Gloucestershire, GL5 4EP

www.amberley-books.com

Copyright © Colin Maggs, 2016, 2017

The right of Colin Maggs to be identified as the Author of this work has been asserted in accordance with the Copyrights, Designs and Patents Act 1988.

ISBN 978 1 4456 7136 9 (paperback)
ISBN 978 1 4456 4097 6 (ebook)

British Library Cataloguing in Publication Data. A catalogue record for this book is available from the British Library.

Typesetting and Origination by Amberley Publishing.
Printed in the UK.

CONTENTS

CONTENTS

I

THE BRUNELS
1806–17

9 April 1806 seemed an inauspicious day when Isambard Kingdom Brunel emerged into a world that was quite unaware of the effect this new arrival would have on land and sea transport, not to mention his profound effect on the appearance of Britain, lasting long after his eventual death.

Who exactly was Isambard Kingdom Brunel? Put simply it could be said that he was an improved version of his father, Marc Brunel, yet Charles Macfarlane, who knew both men, wrote in *Reminiscences of a Literary Life*: 'I had liked the son, but at our very first meeting I could not help feeling that his father far excelled him in originality, unworldliness, genius and taste.'

Marc Isambard Brunel, born on 25 April 1769, was the second son of a tenant farmer at Hacqueville, approximately 28 miles south-east of Rouen. The prosperous family had held the farm for 330 years and enjoyed the privilege of controlling the local posts. As in England, the eldest son was expected to continue with his father's profession while the younger became a priest or a lawyer. Nevertheless, aged eight Marc was sent to the College of Gisors for training as an army officer. He displayed ability at mathematics, drawing and music, becoming proficient at both flute and harpsichord. He used his ingenuity to construct a working model of an instrument which produced the sounds of both. His father still wanted him to become a priest so Marc was sent to the seminary of St Nicaise, Rouen. Although he there displayed an outstanding aptitude in drawing and mathematics, he was less enthusiastic at Greek and Latin. During his free time, Marc enjoyed making a survey of Rouen and sketching its buildings. This eye for detail proved useful. When glancing at a drawing of M. Navier's first suspension bridge across the Seine at Paris, Marc remarked: 'You would not venture, I think, on that bridge unless you would wish to have a dive.' A few days later the bridge collapsed. Later on another occasion, when passing a stone shed at Deptford he said agitatedly to Richard Beamish, 'Quick, quick. Don't you see, it's going to fall!' The next morning it, too, collapsed.

One day when on the quay at Rouen, Marc spotted two large iron cylinders which had just been landed. Enquiries elicited the fact that they were part of a new machine to pump water and that they had been exported from England. 'Ah!' he exclaimed, 'Quand je serais grand, j'irai voir ce pays-la.' In due course this wish came true. When home on holiday he loved being in the village carpenter's shop, but as the Industrial Revolution had yet to influence France, his father, Jean Charles Brunel, continued to force him into the priesthood, not accepting that Marc's future was in engineering. Fortunately the Principal of St Nicaise perceived that Marc's future was not in the Church and managed to convince his father that he should leave. Although Marc gave up Roman Catholicism and cared little for organised religion thereafter, his diaries reveal that he prayed privately. He never forgot his origins and dressed and behaved accordingly. Once when in a British court, he was asked if he was a foreigner. He replied: 'Yes, I am a Norman and Normandy is a country from whence your oldest nobility derive their titles.'

Marc's father now had a problem of what to do with his son. As he was not interested in the Church or the law – while to be a tradesman was quite unthinkable – he would have to be sent away.

Madame Carpentier, an elder cousin of Marc, lived in Rouen with her husband who had resigned as a sea captain to be appointed American Consul in that city. It was decided that Marc could lodge with them, while their friend, Francois Dulagne, Professor of Hydrography at the Royal College, Rouen, could be his tutor, with the view that Marc might be able eventually to enter the navy as an officer cadet. Dulagne was surprised that the propositions of Euclid only required stating once to Marc to be understood without explanation, while after his third lesson in trigonometry he asked permission to measure the height of Rouen's cathedral's spire with an instrument of his own making. In 1786, after a delay due to a smallpox epidemic, the seventeen-year-old Marc joined his ship at Rouen and spent the next six years in the French navy.

On being presented to the captain, Marc noticed a quadrant, an instrument he had never before seen. He examined it without touching, went home and made a reproduction. The first was just a trial, but his second proved so good that he never possessed, or used, another during his service in the navy. Marc had proved himself an expert geometrician, trigonometrician and optician. Starved of music, using his manual skills Marc made a rudimentary keyboard instrument. He was nicknamed Marquis [Marc I]. In January 1792, the ship's crew were paid off at Rouen and for the next eighteen months he again stayed with the Carpentiers.

In January 1793, Marc and Francois Carpentier visited Paris. Only four days before the king's sentence of death was pronounced, Marc, a royalist sympathiser, made a rash public proclamation in a café of the coming doom of Robespierre. Fortunately a friend created a diversion and the two were able to escape from the angry crowd to hide in an inn before creeping out of Paris at nightfall.

It was here at Rouen that he met his future wife, Sophia, daughter of William Kingdom, a Plymouth naval contractor. Following her father's death, her elder brother recklessly sent her to learn French with the Carpentiers. Sophia was the granddaughter of the clockmaker Thomas Mudge, who invented the lever escapement. Shortly after their engagement it was realised that, being a royalist, to escape death Marc must emigrate. Through Carpentier, the American vice consul at Le Havre supplied Marc with an American passport. The revolutionaries tried to restrict emigration, but

Marc was able to leave on the pretext, made plausible by the bread shortage, that he was going to America to buy grain for the navy. En route to the Channel he fell off his horse, whereupon Gaspard Monge, the revolution's navy minister, driving along and completely unaware of Marc's political views, kindly took him to Le Havre where he boarded the *Liberty* for New York, embarking on 7 July 1793. At sea it was soon challenged by a French frigate seeking runaways; Marc then realised his American passport had been lost in his tumble from the horse. Not to be outdone, Marc had two hours to forge a convincing document which succeeded in passing scrutiny – his practical skills had certainly again come in useful. He landed in New York on 6 September 1793 and stayed in the States for six years, gaining a reputation as an architect and engineer.

With two other French emigrants, Marc planned to survey a 220,000-acre tract of land between the 44th parallel, the Black River and Lake Ontario, with a view to parcelling it to other emigrants.

On the way to the area they encountered Thurman, a wealthy New York merchant who agreed to back a plan for a waterway between New York and the Saint Lawrence River. It was when surveying the course for this canal that Marc decided to abandon his intention of rejoining the French navy when the political unrest calmed down and instead become a professional engineer. As the navigation neared completion, Marc entered a competition to design a new capitol building. His proposal was probably based on the Palace of Versailles. Although the judges believed it outstanding, the cost would have been prohibitively high, but New Yorkers loved his scheme and used a simpler version of it for the Park Theatre in the Bowery.

Marc adopted American citizenship and in the autumn of 1796 was offered and accepted the position of chief engineer of New York. Fulfilling this duty he designed, among other things, a cannon foundry, some of the city's earliest civic buildings and advised on the defences of Long Island and Staten Island.

Meanwhile Sophia was still in France. At first she paid for her keep by teaching English, but after the execution of Louis XVI, England joined the alliance against France, so in October 1793 the Council of the Revolution ordered the arrest of all English nationals. Sophia was captured and imprisoned in a convent at Gravelines, midway between Calais and Dieppe. When this temporary prison was full, room was made for newcomers by using the guillotine. On 24 July 1794, the women were freed as Robespierre had fallen. Sophia returned to Rouen where the Carpentiers nursed her back to health.

In 1798, Marc dined with the British aide-de-camp Major-General Hamilton. The table talk reminded Marc of the beam-engine cylinders he had spotted as a lad on the quay at Rouen. It was mentioned in conversation with another French emigrant, Monsieur Delabigarre, that the Royal Navy required 100,000 blocks annually, that Messrs Fox & Taylor of Southampton had secured the monopoly of manufacturing these blocks, and that a seventy-four-gun ship required 1,400 such blocks. This suggested to Marc that he should design better block-making machinery and make a visit to England. Another enticement to go was that he had maintained correspondence with Sophia Kingdom – and indeed had sent her two miniatures: a portrait he had made of her painted from memory and also a self-portrait.

Marc sailed from New York on 7 February 1799, disembarked at Plymouth on 13 March and lodged at Canterbury Place, Lambeth. Within a month of arriving in England, he had designed and patented a polygraph, a device for making up to three copies of correspondence simultaneously.

The Royal Navy required 100,000 rigging blocks annually, all hand made at a total cost of £24,000. Marc designed machinery to do this, but needed two working models to promote his invention. Fortunately a friend advised him to contact Henry Maudslay, one of the very few engineers in the country capable of manufacturing such a machine. The difficulty chiefly lay in getting his idea adopted, as Taylor, the existing manufacturer of the blocks, was not keen to use a new method. General Hamilton, who had met Marc in America, supplied him with a letter of introduction to Earl Spencer, First Lord of the Admiralty, and it was through the earl that Marc gained the friendship of the ruling class. Through Earl Spencer he met Sir Samuel Bentham, Inspector-General of Naval Works. Bentham arranged for Marc to demonstrate his models before the Lords Commissioners of the Admiralty. With Taylor's contract about to expire, Bentham realised the potential value of Marc's design, and on 15 April 1802 recommended the installation of Marc's block machines at Portsmouth Dockyard. Although it took Maudslay six years to perfect the steam-powered block-making machines, these allowed ten unskilled men to do what 110 skilled workmen had done previously. The mass-production era had dawned.

Maudslay was the father of the modern machine shop and his slide-rest lathe, screw-cutting lathe and planing machine opened new standards of accuracy. Shortly after meeting Marc he moved to larger premises which in turn became too small and in 1810 obtained a disused riding school in Lambeth marshes which became the works of Maudslay, Sons & Field.

Marc Brunel and Henry Maudslay made fine partners, Brunel having the ideas and Maudslay having the practical experience to put them into use. His first success was at Portsmouth Royal Dockyard, where they made the wooden pulley blocks. Interestingly, the landing craft which carried troops for the Second World War D-Day landings were equipped with blocks made on Marc's machines.

In the meantime, Marc married Sophia on 1 November 1799 in her parish church of St Andrew's, Holborn. Marc always remembered his wedding anniversary. Aged seventy-six, he wrote: 'To you my *dearest* Sophia I am indebted for all my successes.' Initially they lived in rooms in Bedford Street, Bloomsbury, just a few minutes from Maudslay's workshop; by the end of 1802, Marc no longer needed to be near Maudslay's workshop and moved to a terrace house in Britain Street, Portsea near the dockyard.

Marc's son, Isambard Kingdom Brunel, was born at Portsmouth on 9 April 1806. He had two older sisters, the five-year-old Sophia and the two-year-old Emma. Although Marc Brunel was generally called by his second name, Isambard, to prevent confusion I will refer to him by his first name and refer to his son as Brunel unless it is in a quotation.

Marc invested the money earned from the Admiralty in purchasing a sawmill at Battersea, and so in 1808 moved home from Portsea to 4 Lindsey Row (now corresponding to 98 Cheyne Row), Chelsea. His house was part of a seventeenth-century mansion built for King James's doctor and later rebuilt by King Charles II's Great Chamberlain, Lord Lindsey.

The house stood on the waterfront of what was then a quiet village, with views on one side towards a walled garden and on the other towards the river; trips to London were generally made by boat from steps from the house. Isambard used the proximity of the Thames at the bottom of his garden for swimming and rowing, though as the river was then little more than an open sewer, it is possible that it may have been responsible for weakening his health as he certainly suffered from lengthy illnesses.

Marc won an Admiralty contract for installing timber machinery at Chatham Dockyard and here installed a railway to convey logs to the sawmill. It is significant that the wrought-iron rails were fixed to longitudinal sleepers with occasional cross ties. It is also significant that its gauge was seven feet, so these features may well have influenced Brunel's choice of design for the Great Western Railway's permanent way.

Marc's subsequent inventions were not quite so groundbreaking as the block-making machinery, but carried out such tasks as sawing and bending wood, improving marine steam engines and paddle wheels and designing a machine for making boots for the British Army.

In February 1809, Marc had seen soldiers from the Corunna campaign disembarking and was shocked that many wore not boots, but only filthy rags on their feet. Enquiring, he discovered that although the army spent £150,000 annually on boots they were of poor quality and quickly wore out – unsurprisingly, as clay was placed between the soles to give weight and make them feel stronger than they really were. Marc's machines could make good, strong boots, and the government encouraged him to build a factory to supply the army's entire needs.

Marc accepted this suggestion in good faith and set about manufacturing 400 pairs of boots daily, the machines operated by disabled soldiers – showing a compassionate side to his character. Then, with the peace following the Battle of Waterloo, the army found it had no requirement for the footwear and the vast quantities were left in Marc's hands. To add to this problem, on the night of 30 August 1814 his sawmill at Battersea went up in flames, and, as luck would have it, this coincided with a serious fire on Bankside where most of the fire engines were occupied fighting that outbreak.

Although Marc should have done well financially from his various inventions, the government of the time was typically niggardly. He was unwise in money matters and was too trusting of others; for instance, his partner at the Battersea mill used capital as income. Like the archetypal absent-minded professor, Marc was unworldly and frequently lost umbrellas and caught the wrong stagecoach.

In 1814, Marc was made a Fellow of the Royal Society. That same year he began to be interested in steamships. In 1816, he fitted a double-action engine in *The Regent*, a coastal steamer which ran between Margate and London, making the 90-mile voyage in nine hours twenty minutes.

The Tsar Alexander I, impressed by his work, invited him to St Petersburg. In 1818, Marc designed an 880-foot-long suspension bridge over the River Neva. Funds could not be raised for its construction, so in the same year he designed and patented a cylindrical tunnelling shield to burrow below the river. The idea of the shield was given to him when working in Chatham Dockyard, by watching the humble ship worm, *Teredo navalis*, in ships' hulls. The Russians aborted the Neva tunnel project, but his idea was not wasted as we shall see in due course.

2

YOUTHFUL TRAINING
1817–24

The young Brunel was imaginative, and with his sisters drafted and acted in plays; likewise he was excellent at drawing and painting – attributes which stood him in good stead later in life when designing stations, tunnel mouths and bridges. He enjoyed walking along the top of his garden wall to chatter and joke with his neighbour Miss Mannings.

Marc taught the four-year-old Brunel arithmetic, geometry and how to draw perfect circles freehand, and explained that drawing was the engineer's alphabet. His mother told him tales about her uncle Thomas Mudge who had invented the lever escapement before becoming the king's watchmaker. At the age of eight, Brunel was sent to the Revd Weedon Butler's school in Chelsea where Greek and Latin were added to his curriculum. He took on his father's appearance, being small but with a large head and dark complexion.

Later, Brunel was sent to the Revd Dr Morrell's boarding school at Hove. Dr Morell, a Unitarian, had advanced views for the time, banning bullying, fagging and flogging, even preaching sermons on the advisability of formal education for girls – revolutionary thinking at the time. In addition to Classics, modern languages, geography and mathematics were taught. This school gave Brunel every chance to blossom.

He spent his spare time in Hove just as his father had done at Rouen, surveying the town and making sketches of its buildings, again this proving useful for later work. The eye for detail developed by his artistry made him successfully predict that a building being erected opposite the school was going to collapse – this observation considerably raised his status in the eyes of his fellow pupils. He also carried out boat construction.

In September 1817, aged eleven, Brunel was sent to Marc's nephew, living at some unknown location in France. On 2 September 1817 Marc sent a letter to this French nephew:

I entrust my little boy to you as he needs a Mentor. I don't believe I could make a better choice than in imploring you to moderate the impetuosity of his youth. He is a good little boy but doesn't care for books – except mathematics for which he has a liking. He has started Euclid. Use it as much as you can. You must not let him sit an excessive examination as he often becomes exhausted. As he is not in the best of health he sometimes needs to take a small dose of rhubarb which he carries with him.

He must not be allowed to drink cider.

It was during this visit that Brunel was first introduced to Louis Breguet, a Paris precision engineer. In due course Brunel returned to Dr Morell's school at Hove.

In 1819 Brunel wrote a report to his mother:

I have passed Sallust some time, but I am sorry to say I did not read all, as Doctor Morell wished me, to get into another class. I am at present reading Terence and Horace. I like Horace very much but not as much as Virgil. As to what I am about, I have been making half a dozen boats recently till I have worn my hands to pieces. I have also taken a plan of Hove which is a very amusing job. I should be very much obliged if you would ask Papa (I hope he is well and hearty) whether he would lend me his long measure. It is a long eighty-foot tape; he will know what I mean. I will take care of it, for I want to make a more exact plan, though this is pretty exact, I think. I have also been drawing a little. I intend to take a view of *all* (about five) the principal houses in that great town – Hove. I have already taken one or two.

As Marc believed that mathematics and science were better taught in French schools than elsewhere, after Easter 1820, when he was fourteen, Brunel was sent to the College of Caen in Normandy and in November 1820 to the Lycée Henri-Quatre, Paris, famous for its teaching of mathematics. As formalised engineering degrees were still some way in the future, the most precise engineering experience available was in clockmaking, so Brunel's education was completed by being apprenticed to Louis Breguet, a world-class precision engineer who made chronometers, watches and scientific instruments. On 28 November 1820, Marc wrote to Breguet the following letter:

Monsieur,

Merlin brought me a letter from my son and informed me at the same time of your good disposition towards him. You have already given me proof in his last stay in Paris three years ago when he was still a little schoolboy. I hope that you'll find him more reflective and much less importunate/ vexatious/troublesome. I know that he had got much more inclination towards useful recreation and instructions to go to the theatre or other public amusements and besides, as Sundays are the only days that he can give you, I've told him what your cousin told me on your behalf – that he was to inform you when he did not have any invitation for that particular day.

I would really have liked you to have gone with my family to Paris and to have received your amiable Father and aunt but I was more occupied than ever with the discovery which is connected with a new process of stereotype which Isambard will be able to explain to you with the greatest precision as he has now been my collaborator for several months. I have found him not only very useful but of unflinching perseverance. He never gets tired even though we work after midnight. As a draughtsman he surpasses me and is first class not only by his exactitude with which he copies but he is of a rare frame of mind and he loves work particularly that which holds his attention and involves his hands. He cannot find a better way of occupying himself than by working with you and is persuaded that he will be recognised by it.

I have instructed him to make the best use of your experience and expertise.

Will you accept the assurance of my feelings of gratitude which Madame Brunel and Mme Hawes share with me with all their hearts and with great sincerity?

Marc's financial situation grew serious on account of him paying much greater attention to inventing than accounting, and on 14 May 1821 he was committed to the prison of King's Bench for debt, his wife Sophia insisting on accompanying him to gaol. Marc felt that the government had failed him over the boot episode and, not unnaturally distressed at his situation, wrote to Lord Spencer: 'My affectionate wife and myself are sinking under it, we have neither rest by day or night. Were my enemies at work to effect the ruin of mind and body, they could not do so more effectively.' It was only when Marc thought seriously about taking up the offer to move to Russia, and wrote to Tsar Alexander requesting 'to work under the protection of a sovereign whose enlightenment and liberality seems to shine forth doubly in contrast with the callousness of the government at home', that the British government took action. The Duke of Wellington saw that Marc was granted a sum of £5,000 on condition that he would remain in England. This was sufficient to see him discharged from prison early in August 1821. It is not known who paid for Brunel's education when Marc was in prison. Subsequently, Marc's family tried to avoid referring to this episode in his life and when they did, they referred to it euphemistically as 'The Misfortune'.

In the autumn of 1821, Brunel having finished his studies at Henri-Quatre School, and after a year as an apprentice under Breguet, the young man set his hopes on being admitted to the École Polytechnique. Unfortunately, due to his foreign birth he failed the entrance examination at which 600 French boys competed for less than a hundred places.

On 21 August 1822, Brunel returned to England and worked with his father and solitary clerk in the office at 29 The Poultry. Brunel enjoyed plenty of variety engaging with a range of projects, such as plans for paddle tugs, a cannon-boring machine for the Netherlands government, a new suspension bridge across the Serpentine, swing bridges for Liverpool Docks, and a project with Augustus Pugin for a cemetery at Kensal Green. In fact, such was Brunel's skill that he became Marc's partner, rather than assistant.

Brunel gained valuable experience of suspension bridge design when in 1822 his father's office produced plans for two bridges on the Ile de Bourbon (now Réunion), one possessing a span of 131 feet 9 inches, and the other two spans of this same length. They were designed to withstand hurricanes. Marc was full of wonderful ideas and held patents on such things as expanding rings around pistons to keep them steam-tight and a boiler designed so that when a ship rolled the fire-tubes were never uncovered. This boiler had a dome – the first time such a protuberance had been applied, while the boiler was fired by a mechanical stoker. His smaller inventions included a device for shuffling a pack of cards and a machine for manufacturing nails.

Brunel also spent many happy and profitable hours at Maudslay Sons & Field's world-class engineering works at Lambeth, and with fond recollections, on 24 March 1853, wrote to them saying: 'Your firm with which I gained all my early recollections of engineering and so closely connected and in whose manufactory I probably acquired all my early knowledge of mechanics.'

The high standards of workmanship of Breguet and Maudslay had great influence on Brunel and for the remainder of his life only the very best was good enough for him. This led him into some practical difficulties because he was frequently criticised as extravagant when he insisted on only the finest materials and workmanship.

Apart from Breguet, Brunel was however not impressed by French engineering. He wrote to William Froude, an engineer and applied mathematician, on 7 April 1847:

> Take them for abstract science and study their statistics dynamics geometry etc etc to your heart's content – but never read any of their works on mechanics any more than you would search their modern authors for religious principles. A few hours spent in a blacksmith's and wheelwright's shop will teach you more practical mechanics – read English books for practice. There is little enough to learn in them but you will not have to unlearn that little.

On 23 December 1824, Brunel recorded in his diary:

> I and William [his friend William Hawes] meet three times a week to read the *History of England* by Hume. We read about 50 pages a night, and have just finished the first volume after nine nights' reading.

His youthful education did not consist entirely of work and studying. Brunel and William Hawes jointly owned a 'funny' – a light clinker-built sculling boat employed for competition racing, and used it chiefly when travelling down to Greenwich for a bathe, or to Richmond for dinner.

One such leisurely diversion took place on a Sunday in early September 1825. Brunel, William Hawes, Benjamin Hawes and his wife Sophia (Brunel's sister) left Barge House so early in the morning that they arrived at the White Horse Cellar, Piccadilly by 5.00 a.m., having watched a fire in the Strand on the way. They took the Cheltenham coach to Colnbrook Turnpike, walked four miles to Staines where they breakfasted, and then went off in the 'funny' at about 11.00 a.m. After passing a delightful day, they dined between Kingston and Richmond arriving home by 7.00 p.m.

In 1825 and 1826, Brunel attended the morning lectures at the Royal Institution. The eagerness and rapidity with which he followed the chemical discoveries then being made by Michael Faraday revealed the facility with which he gained and retained scientific knowledge.

3

FIRST ENDEAVOURS
1825–29

In the early nineteeth century, the tidal River Thames formed a barrier between the City of London and its southern suburbs. Ferries were the only way to make a crossing, with some 3,700 persons using them daily. As early as May 1798, Ralph Dodd proposed a 900-yard-long tunnel between Gravesend and Tilbury but the plan proved abortive. Four years later, Robert Vazie proposed a tunnel from Rotherhithe to Limehouse. The Thames Archway Company was formed as a result, the necessary capital raised and a trial shaft sunk. It immediately filled with water and Richard Trevithick was called in to advise. At first Trevithick was appointed resident engineer under Vazie but later became in sole charge when Vazie was dismissed. When the heading was almost complete an abnormally high tide broke in and overpowered the pumps. The Cornish miners carrying out the work along with Trevithick himself narrowly escaped with their lives. The company directors decided that enough was enough and abandoned the works, cutting their losses; William Jessop concluded that 'it was impracticable to make under the Thames a tunnel of useful size by an underground excavation'.

I. W. Tate, one of the Thames Archway promoters, learned of Marc's tunnelling shield, and on 18 February 1824 a meeting was held at the City of London Tavern. The Thames Tunnel Company was established soon thereafter. Marc was to be appointed the engineer at a salary of £1,000 per annum plus £5,000 for the use of his tunnelling shield and a further £5,000 to be granted on completion of the project.

As Rotherhithe was inconveniently far from his home in Chelsea, Marc and his family moved from Lindsey Row to the less fashionable 30 Bridge Street, Blackfriars, set in a smoky, dingy district of the city.

Marc proposed a 1,200-foot-long tunnel three-quarters of a mile upstream from Trevithick's heading, to take carriages beneath the Thames from near St Mary's, Rotherhithe on the south bank across to Wapping on the north bank. The scheme was fraught with risk due to having to bore through clay, mud, sand and gravel. On 22 April 1824, Brunel, aged eighteen, wrote in his diary: 'Took my Theodolite down to Rotherhithe to take the Level of the ground where we are boring.'

Marc used a clever method to obviate the vertical shaft becoming flooded. A brick cylinder fifty feet in diameter and forty-two feet high was built above ground on an iron ring. The next step was to excavate the ground within and beneath so that its weight would cause it to sink until its top reached ground level.

The bricklayers laid 1,000 bricks daily and the shaft reached its full height in three weeks. The ground was then excavated, the shaft, weighing almost 1,000 tons, sinking an average of six inches a day, with crowds coming to watch this unusual sight. On 6 June 1826, the shaft was complete; below tunnel level a reservoir was made to receive water draining from the workings. Steam-driven pumps were employed to lift it out.

At the inaugural ceremony commencing the construction of the tunnel on 2 March 1825, Marc laid the first brick and Brunel the second. Both Marc and his son liked to do things in style. At the subsequent 'sumptuous collation' an imaginative table decoration provided a model of the tunnel created in sugar. The inauguration of the work was a truly public event, with church bells pealing, flags flying, bands playing and streams of famous visitors in attendance such as the Duke of Wellington and Sir Robert Peel. Another interesting detail was that a bottle of wine was ceremonially retained to be opened at a similar banquet when the tunnel was opened – though no one anticipated the number of years it would be until the bottle was at last broached.

On 9 March 1825, Brunel wrote:

> Projects are on foot for Fowey and Padstow Canal, and the Bermondsey Docks. I am preparing plans for South London Docks. In case my father should be named Engineer [of the docks] I am busily engaged on the Gas engine, and the project is likewise made for a canal across the Panama. Surely *one of these* may take place?
>
> It may be curious at some future date to read the state we are in at present – I am most terribly pinched for money. Should receive barely enough next quarter to pay my debts, and am, at the moment, without a penny. We keep neither carriage, nor horse, nor footman, only two maidservants. I am looking forward with great anxiety to this gas engine – building castles in the air about steamboats going 15 miles per hour; going on a tour to Italy; being the first to go to the West Indies, and making a large future; building a house for myself, etc, etc. How much more likely it is that all this will turn out to be nothing! That Gas Engine, if it is good for anything, will only be tolerably good, and perhaps make us spend a good deal of money; that I should pass through life as most other people do, and that I should gradually forget my castles in the air, live in a small house, and at most, keep my gig. On the other side it may be much worse, My father may die, or the [Thames] tunnel may fail.

A few entries from Marc's diaries for 1825 reveal the reliability of Brunel as a twenty-year-old:

> Isambard incessantly in the works, most actively employed, shows much intelligence ... Isambard was the greatest part of the

night in the works ... Isambard was in the frames the whole night
and day until dinner-hour ... Isambard the whole day until 2 at
night ... Isambard has been every night and day too in the works.
I relieved him at 3.

Towards the end of October 1825, Brunel was forced to take three
weeks' rest. On 1 January 1826, Marc Brunel noted regarding the
tunnel:

> There is a considerable degree of *laxity* in the service. The work
> is not carried on regularly. Short of bricklayers in one shift. The
> resident engineer [Armstrong] is too lenient with the men, who do
> as they please. Isambard is the most efficient inspector we have.
> He is constantly at work.

To carry out the tunnelling, Marc's construction shield,
manufactured by Henry Maudslay, was lowered into the vertical
shaft and a hole cut in the brickwork. The shield measured three
men in height and twelve men in width. Each of the thirty-six
men stood in an individual cell-like compartment and dug away
with pick and shovel. Above the heads of the men standing in the
top cells were pivoted plates to support the roof, corresponding
with shoes at the bottom. Plates also protected the sides of the
excavation. Then as the material was excavated the shield was
screwed forward, and bricklayers working back-to-back with
the excavators came in from behind to secure the work. Except
for a space of four and a half inches when the shield was moved
forward, the excavation was continuously supported. Progress
was slow, the average distance being thirteen feet a week. Marc
had desired a driftway below the main tunnel in order to carry off
water, but the directors, thinking to economise, refused. Events
proved it a false saving.

The workers were not to be envied. Both the gas lighting
and ventilation were poor, drainage sometimes left something
to be desired, and there was also the risk of inundation. Poor

ventilation caused sickness and the Thames, which was then little better than an open sewer, also caused widespread illness which claimed a heavier toll than construction accidents; one particularly horrible form of this ailment suddenly stuck men blind. Even if a man recovered – and quite a few did not – he did not always regain his sight. William Armstrong was the resident engineer with Brunel as an assistant. Marc soon became sick with pleurisy.

Marc's diary tells that on 17 March 1826 and the following days, Brunel was engaged on tunnel drawings necessary prior to carrying the work out. On 5 June 1826, Marc wrote:

Isambard got into the drift, and gave the line for a better disposition of the staves, which was afterwards done in a proper manner. Isambard's vigilance and constant attendance were of great benefit. He is in every respect a most useful coadjutor in this undertaking.

And on 8 September 1826:

About 2 p.m. I was informed by Munday that water was running down over No 9 [compartment of the shield]. I went immediately to it. The ground being open, and consequently unsupported, it soon became soft, and settled on the back of the staves, moving down in a stream of diluted silt, which is the most dangerous substance we have to contend with. Some oakum was forced through the joints of the staves, and the water was partly checked. Isambard was the whole night, till three, in the frames. At three I relieved him. He went to rest for a couple of hours; I took some rest on the stage.

On 20 November 1826, Brunel had 'passed seven days out of the last ten in the Tunnel. For nine days on an average of 20 ⅓ hours per day in the tunnel and 3 ⅔ to sleep'.

Marc and Brunel were both intrigued by the idea of a gas, sometimes spelt 'gaz', engine which they hoped could replace steam power and provide cleaner energy. In the spring of 1823, Marc had heard Sir Humphrey Davy read a paper describing Michael Faraday's work on the liquefaction of gases. Gas could be generated from carbonate of ammonia and sulphuric acid and passed into two condensers alternately heated and cooled. The gas in one container was kept in its condensed form by passing cold water through condenser tubes, while the other was heated by the circulation of hot water; the difference in pressure between the two containers was thirty-five atmospheres. It was hoped that this feature could be used to provide a cheaper power than that offered by steam.

Brunel's yearning was shown in his personal diary when he noted:

My ambition or whatever it may be called (it is not the mere wish to be rich) is rather extensive but still am not afraid that I shall be unhappy if I do not reach the rank of hero and Commander in Chief of his Majesty's forces, in the steam (gas) boat department. This is rather a favourite castle in the air of mine. Make the gas engine, fit out some vessels (of course a war), take some prizes nay some island or fortified town, get employed by the government, construct and command a fine fleet of them and fight; in fact, take Algiers or something in that style.

On 30 January 1833, after spending much time and money on the project, Brunel realised that it was time to call a halt to his gas-engine experiments when he wrote:

Gas – after a number of experiments I fear we must come to the conclusion that (with carbonic acid at least) no sufficient advantages on the score of economy of fuel can be obtained. All the time and expense, both enormous, devoted to this thing for nearly 10 years are therefore wasted. It must therefore die and with it all my fine hopes – crash – gone – well, well, it can't be helped.

Brunel certainly fulfilled the family crest of a spur and the motto 'En Avant' for he displayed incessantly this drive to fame. On 14 January 1826, Brunel wrote:

Q – Shall I make a good husband? – Am doubtful. EH [Ellen Hulme] is the oldest and most constant now however gone by. During her reign (nearly seven years!) several inferior ones caught my attention.

In April 1826, Armstrong broke down and resigned from the Thames Tunnel, so the 20-year-old Brunel took over and worked incessantly on the project, sometimes not leaving the tunnel and for five successive days, only pausing to take a brief nap on the wheeled staging behind the shield. He carefully checked the bricks, rejecting any sub-standard loads. The directors, afraid that he would literally work himself to death, provided three assistants: Richard Beamish, William Gravatt and Mr Riley; the latter died on 5 February 1827 after only two months' work.

In an attempt to receive a return from their investment, when 300 feet of the tunnel had been completed the directors decided to admit visitors for a shilling. As the public could not be exposed to the dangerous gas lamps issued to the tunnellers, an oil-gas plant was sanctioned; installation of the pipe and burners began in December 1826. Marc, realising there was a constant risk of flood inundation at the site, was not happy at having these visitors.

In December 1826, the imaginative Brunel arranged a dinner for nine friends in the tunnel.

Brunel was appointed resident engineer on 3 January 1827. On 6 January 1827, when the miners' and bricklayers' wages had been reduced to 2s 10d a day, Brunel recorded:

When pay began this afternoon, the bricklayers refused to receive their wages. They came down to me in the tunnel. They surrounded me, remonstrating, by their spokesman against the reduction and entreating for a continuation of their pay. I thought it time to exercise some authority.

I ordered the men to return to their work and the others to leave the shaft immediately. I was obeyed by both.

When above ground they still persisted in refusing their wages, reiterating their request for the full wages; which could not, of course, be complied with. Isambard continued:

I have every reason to suspect that the miners are in the plot. They, however, have shown no disposition to follow the example. They have received their money, waiting for the end.

On 8 January 1827, one of the portable gas lamps was accidentally dropped down the shaft and exploded. Brunel rushed to the scene, alerted after hearing

... a loud rustling sound, immediately followed by a mass of light aflame all over the bottom of the shaft. It lasted six or eight seconds.

A number of men came running up all terrified, Nelson crying that he was burned all over. I gave him to the care of Gravatt. Robert Greenshield was dreadfully burned. Osborne, Bowling and Davies were also injured more or less. Every attention was paid to the men by applying oil, which relieved them much. The surgeon was sent for.

His father's observations during the period of the accident reflect on Brunel's reaction to the calamity. On 16 January 1827:

Isambard having been up several successive nights, went to bed at ten, and slept till six the next morning. I am very much concerned at his being so unmindful of his health. He may pay dear for it.

Throughout January 1827, the foul air in the tunnel caused an average of 7 per cent of the men to be off sick, including Brunel and Gravatt. On 31 January 1827, Riley became fatally ill.

That Brunel was a highly critical personality is shown by his diary of 12 February 1827: 'Breakfasted after which Gravatt and I dressed to attend Riley's funeral. He certainly was an amiable young man and intelligent but no energy of character and certainly not fit for our work nor likely to have become so.' Beamish subsequently also became ill, lost his sight in one eye and never fully recovered. In order to be instantly available on site, Brunel and Gravatt shared a cabin at Rotherhithe, working twelve-hour shifts for six days a week, frequently working on Sundays.

The partly finished tunnel was first opened to the public on 21 March 1827. It proved most popular and as many as 700 visitors paid a shilling to inspect it. To avoid any problems with intermingling, workmen and visitors used separate staircases.

Brunel stood no ill-discipline in his role. On 9 April 1827, his twenty-first birthday, he wrote:

My birthday. Two steam pumps working hard but pressure will not stop up. When the afternoon shift came on they did not go below but remained on top grumbling about last week's wages [the men had already suffered a pay cut and then Brunel docked a day's pay as a punishment for slow work]. It seemed that the Ganger, Nelson, had not informed them of my intention to make it up to them next Saturday. Pride, who was drunk and seemed very much the cause of it, I immediately turned him out of the yard. Coxon and Stibbs – sober – and grumbling also very much I discharged also, the rest after that were willing to listen and having told them that if they behaved well this week their lost wages should be made up to them next week.

This ploy of withholding pay and keeping it as a hostage for good behaviour was used by Brunel throughout his career. He celebrated his twenty-first birthday by holding a concert in the tunnel.

Watermen informed Marc that his tunnel was nearing a spot where gravel had been dredged and that the shield was likely to break through and the tunnel become flooded. To avoid the problem, on 25 April 1827, Marc hired a diving bell from West India Docks. Lit by candles, it was lowered to the riverbed while air was fed into it from a pump on the support barge. Brunel took his mother and friends down in the bell and indeed discovered the location of the excavation. An iron tube was thrust through the mud into the tunnel workings and a conversation held. Beamish recorded that 'some gold pins supplied by Mr Benjamin Hawes, junior ... were passed up the tube to be presented to friends as a momento of this extraordinary communication'.

On 15 May 1827 Brunel noted:

Nothwithstanding every prudence on our part, a disaster may still occur. *May it not be when the arch is full of visitors!* It is too awful to think of it. I have done my part by recommending to the directors to shut the tunnel.

On 17 May 1827, nine coal boats anchored above the shield and in the evening Brunel examined the bed and found it secure. That day he wrote:

The water increased very much at 9 o'clock. This is very *inquiétant*! My apprehensions are not groundless, I apprehend nothing, however, as to the safety of the men, but first the visitors, and next, a total invasion of the river. We must be prepared for the Worst.

On 18 May 1827, Marc Brunel and Richard Beamish escorted Lady Raffles and a party of her friends around the tunnel. They left at 6.00 p.m., Brunel and Gravatt going above ground leaving Beamish in charge. The latter wrote:

My holiday coat was exchanged for a strong waterproof, the polished Wellingtons for greased mud boots, and the shining beaver for a large-brimmed south-wester.

The tide was now rising fast. On entering the frames, Nos. 9 and 11 were about to be worked down. Already had the top polings [planks which held up the tunnel face; one was removed from each box where a miner was busy working to allow him a gap of four and a half inches to shovel out the spoil] of No. 11 had been removed. Goodwin, a powerful and experienced man, called for help. For him to have required help was sufficient to indicate danger. I immediately directed an equally powerful man, Rogers, in No. 9, to go to Goodwin's assistance; but before he had time to obey the order, there poured in such an overwhelming flood of slush and water, that they were both driven out, and the bricklayer [William Corps] who had also answered to the call for help, was literally rolled over on to the stage behind the frames, as though he had come through a mill sluice, and would have been hurled to the ground, if I had not fortunately arrested his progress.

I then made an effort to re-enter the frames, calling upon the miners to follow; but I was only answered by a roar of water, which long continued to resound in my ears. Finding that no gravel appeared, I saw that the case was hopeless. To get all the men out of the shield was now my anxiety. This accomplished, I stood for a moment on the stage, unwilling to fly, yet incapable to resist the torrent which momentarily increased in magnitude and velocity, till Rogers, who alone remained, kindly drew me by the arm, and, pointing to the rising water beneath, showed only too plainly the folly of delay. Then ordering Rogers to the ladder, I slowly followed.

As a singular coincidence, I may here remark that this man, Rogers, who showed such kindly feeling and devotion, had served with me in the Coldstream Guards.

Beamish and Rogers struggled through the waist-high water to the barrier which prevented the visitors proceeding further into the tunnel.

Beamish continued:

Arrived at the barrier, four powerful hands seized me, and in a moment placed me on the other side. On we now sped. At the bottom of the shaft we met Isambard Brunel and Mr Gravatt. We turned. The spectacle which presented itself will not readily be forgotten. The water came on in a great wave, everything on its surface becoming the more distinctly visible as the light from the gas-lamps were more strongly reflected. Presently a loud crash was heard. A small office, which had been erected under the arch, about a hundred feet from the frames, had burst. The pent air rushed out; the lights were suddenly extinguished, and the noble work, which only a few short hours before had commanded the homage of an admiring public, was consigned to darkness and solitude.

For the first time I now felt something like fear, as I dreaded the recoil of the wave from the circular wall of the shaft, which, if it had caught us, would inevitably have swept us back under the arch. With the utmost difficulty, the lowest flight of steps was cleared, when, as I had apprehended, the recoil came, and the water surged just under our feet. The men now hurried up the stairs and, though nearly exhausted, I was able to reach the top.

The last man had scarcely reached the top of the spiral stairs when the bottom flight was swept away. Then a faint cry for help was heard from the bottom of the shaft. It was Tillett, the old engineman who had gone down to repack his pumps.

Beamish continued writing:

I was looking about me how to get down, when I saw Brunel descending by a rope to his assistance. I got hold of one of the iron ties, and slid down into the water hand over hand with a small rope, and tried to make it fast round his middle, whilst Brunel was doing the same. Having done it, he called out, 'Haul up'. The man was hauled up. I swam about to see where to land. The shaft was full of casks. At a roll call no one was missing, Beamish mounted a pony and rode to Bridge Street.

Brunel was not a man to sit down and feel sorry for himself. Within twenty-four hours he had borrowed a diving bell from the West India Dock Company to investigate the problem. He was able to stand with one foot on the end of the tunnel brickwork and the other on the tail of the shield, a feat he recorded in a little graphic drawing. The problem of the inundation was rectified with bags of clay, strengthened with hazel rods, thrown into the depression and sealed with a raft loaded with 150 tons of clay. The pumps began to lower the water level, but then the following day it rose again – the ebb tide had caused the raft to tilt.

Marc, who had been persuaded to adopt the raft plan, decided to raise the raft and rely only on the clay bags. To form a secure bed for the bags, iron rods were laid over the space between the shield and brickwork. To superintend this operation, Brunel descended in the bell when Beamish, standing on the barge supporting the bell, was shocked to observe the floor of the bell suddenly float to the surface. Then, much to his relief he felt a tug on the communicating cord and hauled the bell in. Pinckney, the assistant down with Brunel, had stepped out of the bell on to ground which collapsed beneath him. Brunel, thinking quickly, held out his foot for him to grab, and in the struggle the floor footboard became detached. He hauled Pinckney into the bell and both were brought safely to the surface. This episode failed to faze Brunel, who continued further descents in the bell.

By 11 June 1827, the pumps had emptied the shaft; 150 feet of the tunnel was cleared by 25 June. On 27 June, Brunel decided to inspect the tunnel by boat – a highly dangerous operation because had the river broken in again there would have been no hope of escape, there being only just sufficient room for the boat to move along under the tunnel roof. Throughout his life he enjoyed living dangerously. Dressed in bathing trunks and carrying bullseye lanterns they glided down the tunnel, watched by a group assembled at the foot of the shaft. Part of the tunnel was so deep in water that they propelled the boat by thrusting against the tunnel roof. They were relieved to see that the shield, although filled with silt, was undamaged. Brunel called for three cheers and three more cheers were returned from those at the foot of the shaft.

Clearing the stinking silt from the tunnel was unpleasant and caused the men again to become giddy and sick. Marc wrote on 7 July 1827:

Very uncomfortable in the frames; the candles cannot burn, the ventilator cannot act. Can there be a more anxious situation than that which I am constantly in? Not one moment of rest, either of mind or body. Beamish always ready. Poor Isambard always at his post, too, alternately below or in the barges, or in the Diving Bell.

Even though the tunnel was still partly flooded and dangers were in abundance, visitors still wished to inspect the works. On 17 June 1827, two such were Charles Bonaparte and the eminent geologist Sir Roderick Murchison. The latter recorded:

The first operation we underwent (one which I never repeated) was to go down in a diving-bell upon the cavity by which the Thames had broken in. Buckland and Featherstonehaugh, having been the first to volunteer, came up with such red faces and such staring eyes, that I felt no great inclination to follow their example, particularly as Charles Bonaparte was most anxious to avoid the dilemma, excusing himself by saying that his family were very short-necked and subject to apoplexy, etc.; but it would not do to show the white feather; I got in, and induced him to follow me. The effect was, as I expected, most oppressive, and then on the bottom what did we see but dirty gravel and mud, from which I brought up a fragment of one of Hunt's blacking bottles. We soon pulled the string and were delighted to breathe fresh air.

The first folly was, however, quite overpowered by the next. We went down the shaft on the south bank, and got, with young Brunel, into a punt which he was to steer into the tunnel till we reached the repairing shield. About eleven feet of water were still in the tunnel, leaving just enough space above our heads for Brunel to stand up and claw the ceiling and sides to impel us. As we were proceeding he called out, 'Now, gentlemen, if by accident there should be a rush of water, I shall turn the punt over to prevent you being jammed against the roof, and we shall then be carried out and up the shaft!' On this C. Bonaparte remarked, 'But I cannot swim!' And, just as he had said the words, Brunel, swinging carelessly from right to left, fell overboard, and out went the candles with which he was lighting up the place. Taking this for the *sauve qui peut,* fat C.B., then the very image of Napoleon at St Helena, was about to roll after him, when I held him fast, and, by the glimmering light from the entrance, we found young Brunel, who swam like a fish, coming up on the other side of the punt, and soon got him on board. We of course called out for an immediate retreat, for really there could not be a more foolhardy and ridiculous risk of our lives, inasmuch it was just the moment of trial as to whether the Thames would make further inroad or not.

The foolishness of such exploits became clear ten days later when two of the tunnel company directors insisted on inspecting the situation. Accompanied by Brunel's assistant, Gravatt, and two miners, they had reached deep water when one director stood up, struck his head on the roof of the tunnel and fell backwards, capsizing the punt. The only swimmers present were Gravatt and one of the miners; these two swam back to the shaft and collected another boat. They found the two directors clinging to the side of the punt, but there was no sign of the other miner. Unfortunately he had drowned – the first person to die by accident in the tunnel.

When the water level fell, the shield had first to be cleared of silt; there was always present the danger of a further inundation. The miners in the shield were not easily worried, but the superstitious Irish labourers believed that if the lights were put out the water could not find them.

In August 1827, Marc had become seriously ill and Beamish suffered an attack of pleurisy and was absent for six weeks. The inundation had interfered with the tunnel's ventilation system and the foul air formed a black deposit around the men's nostrils. They often collapsed with violent attacks of giddiness and vomiting.

By November 1827 the shield had been completely cleared and proper work could be restarted.

On 7 October 1827, walking in the dark from his office to the shaft Brunel fell into an uncovered water tank. Although quite seriously hurt he refused to go home and declined to remove his boots to inspect the damage as he feared that, if he took them off, he would be unable to replace them should an emergency arise. At 2.00 a.m. on 8 October he finally assented to Gravatt removing his boots and sending him home in a cab. 'I went to bed. My knee remains very swollen ... felt sick and shaken. Sent for Doctor Bardell, leeches etc. etc.' Brunel did not return to work until 24 October.

It was while convalescing from his fall into the tank that he had time to contemplate his love life. Certainly by 1827, when Brunel was twenty-one, he had thoughts of marriage, but questioned whether it would be a right move. He wrote of apprehensions in his private diary: 'Shall I make a good husband – am doubtful – my ambition, or whatever it may be called (it is not a mere wish to be rich) is rather extensive.' 'My profession is after all my only fit wife.' 'I have always wished and intended to be married, but I have been very doubtful on the subject of – children – it is a question whether they are the sources of most pleasure or pain.'

Brunel had been trained by his father to keep a diary, but in October 1827 he felt the need for a journal in which he could record intimate thoughts and insights, so to this end he obtained a small leather-bound notebook secured by lock and key. In his new private journal Brunel recorded excitedly on 13 October 1827:

I wish I had kept this journal with me even at work on the river. What a dream it now appears to me! Going down in the diving bell, finding and examining the hole! Standing on the corner of No. 12! [i.e. the twelfth chamber of the shield]. The novelty of the thing, the excitement of the occasional risk attending our submarine (aquatic) excursions, the crowds of boats to witness our works all amused – the anxious watching of the shaft – seeing it full of water, rising and falling with the tide with the most provoking regularity – at last, by dint of clay bags, clay and gravel, a perceptible difference. We then began pumping, at last reaching the crown of the arch – what sensations!

I must make some little Indian ink sketches of our boat excursions to the frames; the low, dark, gloomy, cold arch; the heap of earth almost up to the crown, hiding the frames and rendering it quite uncertain what state they were in and what might happen; the hollow rushing of water; the total darkness of all around rendered distinct by the glimmering light of a candle or two, carried by ourselves; crawling along the bank of earth, a dark recess at the end – quite dark – water rushing from it in such quantities as to render it uncertain whether the ground was secure; at last reaching the frames – choked up to the middle rail of the top box – frames evidently leaning back and sideways considerably – staves in curious directions, bags and chisel rods protruding in all directions; reaching No. 12, the bags apparently without support and swelling into the frame threaten every minute to close inside brickwork. All bags – a cavern, *huge, misshapen* with water – a cataract coming from it – candles going out.

Altho' to others I appear in such cases rather unconcerned and not affected (pride) the internal anguish I felt is not to be described. I thank God however I rather returned thanks that it was no worse than grumbled. I am an optimist I hope in deed as well as word. 11½ pm – I will le down on the rug a little and then go below.

The final entry in this short-lived private diary was on 6 April 1829 when he seemed to believe he was destined for an early death.

Diary entries were in four groups, the first covering October to December 1827, dominated by the Thames Tunnel and relations with the female sex. The second group covered April to August 1828, mentioning the inundation of the tunnel on 12 January 1828 when he almost lost his life. Several of the entries at this period were written in bed; matrimonial worries were also penned. The third group consisted of just one rather despondent entry on 6 April 1829, while the fourth group was a single entry on 8 April 1829, the eve of his twenty-third birthday. Probably due to its being a very personal document rather than for public consumption, Brunel's punctuation, syntax and spelling are not always conventional.

His first entry, written on 11 October 1827, reads:

At last I have begun this my private journal even now altho' at the second line I can hardly perswade [*sic*] myself that it is really private but am puzzling myself for *proper* words thus destroying the very object I have in view viz to record my feeling habits faults *wishes* hopes and everything belonging to the present moment. The pleasure I shall derive hereafter in reading and comparing the remarks made at different time will I promise myself be very great. I think also with good will much utility and some good lessons may be go. To begin.

My present intentions are to put down whatever is uppermost in my mind without order or arrangement.

I have had this book 8 or 10 weeks: have had many better opportunities, why have I not begun before? I have postponed it! I am very prone to say tomorrow and yet I am not irresolute nor do I want firmness in greater matters. I must cure myself of this – *tomorrow I will begin.*

This is Saturday night 12½ o'clock. A few hands at work below. Williams just gone home, sitting by my fire writing this – it's a great Luxury is being alone – and comfortable. My life has been as yet pretty full of varieties: what shall I be hereafter? At present thinking of nothing but the *tunnel:* living here entirely at Rotherhithe I get up every morning at 8 a.m., see the various things concerning the works all the morning. We dine at 3 p.m., one week boarding with me then a week at Gravatts [*sic*] and so on. After diner [*sic*] I come on duty more particularly taking charge of the works below till 2 in the morning when Gravatts [*sic*] gets up and we sup together. I then go to bed on the sofa in the parlor [*sic*] and Gravatt comes on duty – tired already – in fact am very sleepy – about 1 a.m. – will go to bed.

The 14 October 1827 entry in his private diary read:

> Sunday night – 10 p.m. West parlour Rotherhithe. A new
> broom sweeps clean! The second night and again at it! I have
> been today to Mr Sweets [Marc Brunel's solicitor] about
> agreement but did not see him.
>
> When thinking of this journal whole volumes crowded
> on me – my character my '*châteaux d'espagne*' &c – now
> I am quite without an idea. As to my character – N.B. I am
> a phrenologist. My self conceit and love of glory or rather
> approbation vie with each other which shall govern me. The
> latter is so strong that even of a dark night riding home when
> I pass some unknown person who perhaps does not even look
> at me I catch myself trying to look big on my little pony. The
> former upon reflection does not seem too strong to counteract
> the latter. I often do the most silly useless things to appear to
> advantage before or attract the attention of those I shall never
> again see or whom I care nothing about. The former renders
> me domineering intolerant nay even quarrelsome with those
> who do not flatter me in this case; both combine to make me
> unpleasant at home.
>
> [To] build a splendid Manufactory for gaz engines, a yard
> for building the boats for Do. &c. at last be rich have a house
> built of which I have even made the *drawings* &c. Be the first
> Engineer and an example for future ones.

Brunel's journal continues to recollect another incident:

> October 17th. 2 a.m. Having seen all right below I retired as
> usual to supper; just as I had taken off my coat and boots,
> Kemble [the overground watchman] came in rather a hurry
> to tell me in a trembling voice that the water was in again;
> I could not believe him – whilst hastily pulling on my boots
> I asked him how it happened – how he knew it. 'It was up
> to the shaft when I came, Sir.' I ran without my coat as fast
> as I could giving a double knock on Gravatt's door on my
> way. I saw all the men on top of the shaft, looking for, and
> anxiously calling to those who they fancied had not had time

to escape. Indeed, Miles had already in his zeal for the safety of others, thrown a long rope, swinging it about, calling on the unfortunate sufferers to lay hold of it, encouraging and cheering those who could not find it to swim to one of the landings. I ran, or rather threw myself, down the staircase. The shaft was completely dark. I expected at every step to splash into the water.

So quickly did I go, that before I was at all aware of it I found myself in the tunnel. I met one or two running out. I soon reached the frames in the eastern arch, and saw Pamphilion, who told me that *nothing was the matter, but a small run in number 1;* found Huggins and the *corpe d' élite* there, and a run of mud at the west to front corner – nothing serious. One of the bricklayers' labourers, hearing the man in number one call for assistance and straw, thrust in a whole truss of straw, and then, overcome with fear, jumped right off the stage crying, 'Run! Run! Murder! Put the lights out!', and his fellow labourers, following like sheep, repeated these extraordinary cries. Most of the men being at supper, and distant from, the frames, the panic spread rapidly.

Miles' dramatic version of the affair was:

I seed them there Hirishers a come a tumblin' through one o' them small harches like mad bulls, as if the devil picked 'um – screach of Murther! Murther! Run for your lives! Out the lights! My ears got a singin', Sir – all the world like when you and me were down in that 'ere diving-bell – till I thought as the water was close upon me. Run legs or perish body! Says I when I see Pascoe ahead o' them there miners coming along as if the devil was looking for 'im. Not the first, my lad, says I, and away with me – and never stopped till I got landed fair above ground.

Isambard's personal journal later records on 5 November 1827:

Am just recovered from a serious bruize [*sic*] – fell into the river water well and hurt my knee. Intend giving a dinner next

Saturday in commemoration of or as a thanksgiving for our
having completed 20 ft since the grand water battle – which
we shall have done please God at that day.

Brunel felt urged to celebrate his return to work with a spectacular
banquet, to be actually held in the tunnel itself; thus on 10
November 1827 fifty guests dined at a long table lit by four
decorative candelabra from the Portable Gas Company, mounted
on plinths flanking the table. Those attending had their ears
tickled by the band of the Coldstream Guards who played,
among other things, the triumphant 'See the Conquering Hero
Comes'. The connecting arches in the dividing wall were hung
with crimson drapes.

Admiral Sir Edward Codrington was one of the tunnel's
promoters and the friend who had supported Marc when
he was in the debtors' prison, so when a toast to his health
was proposed, one of his friends quoted from that evening's
Gazette Extraordinary Codrington's victory over the Turkish
fleet, which led to the liberation of Greece from Turkish rule.
Brunel's friend declared: 'In that battle the Turkish power has
received a severer check that it has ever suffered since Mahomed
drew the sword. It may be said that the wine-abduring Prophet
conquered by water – upon the element his successors have
now been signally defeated. My motto, therefore, on this
occasion, when we meet to celebrate the expulsion of the river
from this spot is "Down with water and Mahomed – wine and
Codrington forever!"' The health of the Royal Engineers, the
civil engineers and other good people were also drunk. Marc
Brunel, by his absence from the banquet, projected Isambard as
his de facto successor.

Brunel himself did not overlook the workmen on this auspicious
occasion – 150 of them dined under a nearby arch.

On 12 November 1827, Brunel noted: 'Portable gaz man here to
fetch away things; although some were damaged, he was extremely
civil.'

In his personal journal he penned:

Wednesday evening November 21st 1827.

Returned from Town. Put on my fighting dress having given some directions at Smiths shop screwing gauge for new long jacks &c &c having in fact smelt the cold, frosty, night air. On entering my parlour a nice blazing fire, my table before it, papers, books &c nicely arranged thereon. The whole inspired such a feeling of comfort that I could not resist sitting down and imparting my sensations to this book as to a friend.

As long as health continues, one's prospects tolerable and *present* efforts, whatever they may be, tolerably successful, then indeed a bachelor's life is luxurious. Fond as I am of society *'selfish comfort'* is delightful. I have always found it so. My *chateaux d'espagne* have mostly been founded on this feeling. What independence! For one who's [*sic*] ambition is to distinguish himself in the eyes of the public such a freedom is almost indispensable. But on the other hand, in sickness or disappointment how delightfull [*sic*] to have a companion whose sympathy one is sure of possessing her dependence on you gives her power to support you by consolation and however delightfull [*sic*] may be the freedom and independence, still we find that certain restraints are necessary for the enjoyment of pleasure. I have always wished and intended to be married but have been very doubtfull [*sic*] on the subject of children. It is a question whether they are the source of most pleasure or pain. I have had, as I suppose most young men must have had, numerous *attachments*. If they deserve the name each in its turn has appeared to me *the true one*. E.H. [believed to be Ellen Hulme of Manchester] is the oldest and most constant, now however gone by. During her reign, nearly 7 years, several inferior ones caught my attention. I need only remind myself of Mlle D.C., O.S., and numerous others. With E. H-e it was mutual and I trust the present feeling is also mutual in this case the sofa scenes &c must now appear to her, as to me, rather ridiculous. She was a nice Girl and had she improved as a girl of her age ought to have done ...[the remainder of the page has been cut out and continues]… have served me right if I had been spill'd in the mud. Certainly a devilish pretty girl, an excellent musician and a very sweet voice. But I am afraid those eyes don't speak of a very *placid temper.*

I have nothing after all so very transcendent as to enable me to rise by my own merit without some such help as the Tunnel. It's a gloomy perspective and yet bad as it is I cannot with all my efforts work myself up to be down hearted. Well, it's very fortunate I am so easily pleased. After all let the worst happen – unemployed, untalked of, penniless (that's damned awkward) I think I may depend upon a home at Benjamin's. My poor father would hardly survive the tunnel. My mother would follow him. I should be left alone – here my invention fails, what would follow I cannot guess.

Time proved that E. H. had not 'gone by' for on 8 June 1828 he wrote:

I have not opened this book for a long time to record my old attachment. Who would have thought I should ever have lost anything from over-modesty? It appears that I really might have had A B …w [unidentified] a fine girl, plenty of accomplishments and £25,000 – no joke. S. R…s [unidentified] assures me of it. W W…s [William Hawes?] saw it and B…sh [Richard Beamish] also. Well it's all for the best again. It would never have done to have married then – quite absurd – so young and when it came to the point I should have found too late what I now find – every day I return to my old love – Ellen is still it seems my real love. I have written her a long letter yesterday. Her answer shall decide. If she wavers, I *ought* to break it off for I cannot hope to be in a condition to marry her and to continue in this state of suspense is wronging her. After all I shall most likely remain a batchelor [*sic*] and that is I think best for me. My profession is after all my only fit wife.

Oh that I may find her faithfull [*sic*] and an honor [*sic*] to me. How time and events creep on. Next Wednesday is the public meeting. Shall we get the money? To be or not to be? If we don't – why. And the [tunnel] works suspended it *can* only be temporary they will be secured and opened as a show and in the meantime we must try and get the government to begin borings &c at

Gravesend. This would after all not be so bad – Oh Ellen Ellen if you had kept up your musick and can even play tolerably we might be very happy yet. And starve. It won't do. However we'll see. If the Tunnel stops our main hopes must be on Gravesend. My father's ideas of going abroad for a time will never do. To lose all the other business, Oxford Canal and sundries &c would alone support us and Gravesend with what little salary I get from the Tunnel for they must have a resident Engineer will enable us to *live*.

If I had married A B…w [unidentified] I should certainly have been independent to a certain degree tho' I should hardly have liked depending on my wife. She'd have made a good one tho', I think. But it would spoil all my future prospects I'm afraid to *settle* so early.

Oh for a lighthouse: I must find some place where one is wanted, besides my scheme which I really think a good one, the carrying on such a work exposed to sea storms and the devil knows what would just have suited me. If we can get any body to go on with the gaz machine. Oh dear many many irons and none hot! If it was not for B-n [Benjamin Hawes] I really should be blue dear fellow he's a good fellow and has but one fault and that's more than I can say for any other being I know. Well I swear I'll never take offence at him whatever might happen and then we may live and die friends. Damn it Ellen how you keep creeping on me here I am thinking of you again – Well until I have your answer I cannot fix my mind on the subject and therefore it is no use thinking of you.

My father is gone to Towcester. Gravatt in the country. Beamish working away at the Tunnel. I still in Bridge St, working at Mathematics. Intend going to Redriff [Rotherhithe] next week. Am now almost *quite better* as they say. After nearly 5 months spell pretty well I think.

The romantic intrigue continued when, on 13 June 1828, Brunel noted in his personal journal:

No answer yet from E…n [Ellen] and I'm afraid when it comes it will be a quizzing one without any decisive answer. A shocking habit that of quizzing it at last prevents a person from thinking seriously. I'm almost afraid of an answer however for to marry would be absurde [*sic*] and to remain for years engaged would be painfull [*sic*]. I fear that I ought not to marry unless I find a wife with a fortune.

A Batchelor's [*sic*] life is a very happy one I shall have every advantage of introduction into the best society and a wife unless she is of that class in her own right becomes a bar.

As a Batchelor I may be sure of living tolerably at my ease while with a wife I have every prospect of starving.

If I have anything like an answer it will probably decide my state for life.

And on 17 August 1828, in a philosophical mood:

The last time I wrote I appeared to be thinking of nothing but my answer from Ellen. We are now in correspondence but I do not putting the thing home. I am half afraid of my old attachment binding me and yet I have not the heart to break it off.

The Tunnel is now *blocked up* at the end and all work about to cease. A year ago I should have thought this terrible and not to be born [*sic*]. Now it is come, like all other events it is only at a distance that they appear to be dreaded. Time present seems to me *allways* [*sic*] *alike* to a person who only looks at the future or the past the present situation is quite disregarded like a traveller enjoying a beautyfull [*sic*] landscape or admiring a fine view the mere spot on which he stands never enters into the picture. If the prospect before him is more dismal than that through which he has passed he looks behind him. He may thus always [find] something to admire. Just the moment that the tunnel ceases to be a resource the arbitration of the Battersea concern is likely to be terminated and thus most opportunely supply ways and means. I have always found it so. Either we are peculiarly favored [*sic*] or else misfortune must consist more in discontent than in reality.

The poor Hulmes are in a very unpleasant situation [probably financial]. I wish I could go down and see them indeed I must go soon for other reasons.

Beamish is gone to Ireland: adieu to him. We shall soon all break up. I shall recommence our work on the Gas. I am rather anxious about the specific heat of the liquid: now if this also should fail, down go a good percentage of my Castles in the air. Well can't help it. I wish only that I was the only one concerned.

The love for some unknown reason dissipated, for eight months later on 6 April 1829 he wrote in his personal journal: 'I have had long correspondence with Ellen which I think I have managed well. I may now consider myself independent.'

Ellen was seemingly a lively girl who enjoyed avoiding giving direct answers to questions. This would have made Brunel feel uncomfortable through not being in charge of the conversation or situation. Later information about Ellen is scarce, but it is believed that she remained a spinster because in his private letter book of 20 January 1858 he refers to an annuity paid to Miss E. Hulme and her sister:

> The arrangement I had wished to carry out was to have two separate annuities of £50 each and the balance invested in one on the survivor of the two; so that while they were both living they would get the three annuities amounting to £230 and I think that the survivor would get £180, which I thought a wise proposition.

On 2 January 1828, some of the rocks used to fill the depression in the riverbed fell into the tunnel workings. Overconfident that the shield would pass the point and the brick arch support the riverbed, Brunel failed to dump more clay into the river.

Work on the Thames Tunnel continued and high-ranking visitors came to inspect the works, including Don Miguel, Pretender to the throne of Portugal on 8 January 1828. The following day,

9 January 1828, Francis Giles, the canal engineer who had mapped the Thames riverbed in 1824, visited the tunnel, but despite the hospitality of Beamish and Brunel, left before Gravatt could advance a frame. Since silt was coming into the workings, this was quite understandable.

Isambard had a narrow escape only three days later. He had come on duty just before 6.00 a.m. and issued punch made from gin and warm beer to the miners, who had had a rather unpleasant night. Brunel was standing in the back of the box and watched while the two top polings were cut back and replaced. Suddenly, clay followed by a torrent of water burst through the space between the second and third boards. Brunel realised that the breach was irreparable and ordered all men out of the tunnel. By this time the water had doused the lights and Ball and Collins, followed shortly by Brunel, struggled through the darkness waist-deep in water. They were feeling their way along the stage when it suddenly leaned over and pinned them down beneath the water. Brunel managed to struggle free and called on Ball and Collins to follow him to the shaft. Brunel later described the chaotic scene:

> I struggled under water for some time and at length extricated myself from the stage, and by swimming and by being forced with the water, I gained the eastern arch where I got a better footing, and was enabled, by laying hold of the railway rope, to pause a little in the hope of encouraging the men who had been knocked down at the same time as myself. This I endeavoured to do by calling them. Before I had reached the shaft, the water had risen so rapidly that I was out of my depth, and there swam to the visitors' stairs, the stairs for the workmen being occupied by those who had so far escaped. My knee was so injured by the timber stage that I could scarcely swim.

A watchman burst into the office at the top of the shaft yelling: 'The water is in! The Tunnel is full!' As the workmen's stairs were blocked by escaping men, Richard Beamish, using a crowbar, opened the door of the visitors' steps. He had only gone down a few steps when a wave of water surged up the shaft and on it was the still form of Brunel whom he was able to grab; had Beamish been only a few seconds later, the wave would have receded and swept Brunel out of reach and thus to his certain death. Beamish carried him up the stairs and laid him on the ground. He kept murmuring, 'Ball, Collins.'

Brunel had had a wonderful escape: if Fitzgerald had not opened the visitors' door with its unfamiliar catch, and if Beamish had not caught him, Brunel would have drowned.

Unfortunately, Thomas Ball died, leaving a widow; John Collins left a wife and two children; John Long left a widow; Jeptha Cook left six orphaned children; while Thomas Evans and William Seton were single men who drowned.

A dramatic report appeared in a paper:

Wives and children in a state of nudity, the accident happening at such an early hour, were seen in the utmost state of distress, eagerly enquiring after their husbands and fathers.

In the midst of this distressing scene, out rushed a number of men in a state of great exhaustion, some carrying their fainting comrades, who were removed to the house of Mr Timothy, the Spreadeagle, where they were brought to by restoratives.

Although Brunel was far from well, he insisted on remaining at the works and ordered the diving bell barge and diving bell to be prepared. Unable to walk, he directed operations lying on a mattress on the barge deck and did not return home until 2.30 a.m. on 12 January 1828, when he learned of the size of the cavity in the riverbed. 4,500 tons of clay had to be dumped before the tunnel could be begun to be cleared of water.

It was indeed very fortunate for Gravatt that Brunel with his expertise was on the barge and could direct his descent in the bell. The lowering chain was too short and Gravatt could only probe the riverbed with a rod. Being Sunday, a longer chain could not be obtained but a cable was hastily found. Brunel forbade the use of this – and it was lucky that he did, for while the bell was being hauled up after an unmanned descent, this rope snapped.

Dr Morris transferred Brunel to Barge House in his carriage, Brunel writing:

> Never felt so queer, could not bear the least shake, felt as if I should be broken to pieces; put to bed, was cut, I believe, that night, but don't remember.

The *New Times* reporter recorded on 15 January 1828:

> It is gratifying to state that Mr Brunel junior is rapidly recovering from the effects of his recent accident; and it is much to the credit of the faculty, and highly praiseworthy to the individuals as men, to state that many physicians of eminence and popularity have spontaneously visited Mr Brunel for the purpose of offering any assistance which their talents might afford. The state of the weather and the nature of his complaint have imperatively constrained him to keep to his apartment the whole of yesterday; and Mrs Brunel ... has at present taken up her residence in the same house with her son, and exercises that maternal attention to him which the situation requires. The elder Mr Brunel, so far from giving way to that despondency, which some misdirected accounts have attributed to him, appears to possess additional fortitude and determination.

The *Mechanics' Magazine* had criticised Brunel for not taking advice or accepting responsibility for his mistakes. Fortunately, the tunnel directors supported Brunel, publishing in *The Times, New Times, Herald, Leger* and *Courier* the notice:

> That this Court, having learned with great admiration of the intrepid courage and presence of mind displayed by Mr Isambard Brunel, the Company's resident engineer, when the Thames broke into the Tunnel on the morning of the 12th instant, are desirous to give their public testimony to his calm and energetic endeavours, and to that generous principle which induced him to put his own

life in more imminent hazard to save the lives of the men under his immediate care.

As part of his convalescence, on 27 January 1828 Brunel travelled to Brighton with his friend Benjamin Hawes, staying at the Albion. Hawes returned to London the following evening, but Brunel enjoyed himself exploring the town, going to the theatre, dining and attending a fancy dress ball.

He wrote:

Monday 4th February. Very comfortable at the Albion – some pleasant company – Strolled on the pier smoking my Meerschaum before breakfast – breakfast at twelve. Rode about – visiting *works* (sea-wall being built). Dined at six or seven. Very busy doing nothing all day.

On 8 February 1828, he was less cheerful, suffering a violent haemorrhage after riding a horse. A doctor was sent for and Brunel put to bed. On 11 February he was up but still bleeding, and 15 February saw him 'still ill and retrieved by Beamish in a hired chariot which brought on more bleeding'.

Brunel intended starting work, but when back at Blackfriars found he needed much longer to recuperate; in fact he was so seriously injured internally that he was off sick for over three months.

Rest seemed to have improved Brunel's health so much that on 15 March 1828 he enjoyed a boat trip to Chelsea with his mother, but this venture was too premature and yet another haemorrhage followed. Two new doctors, Travers and Brown, were called, Brunel recording: '23 March. Mr Travers bled me, and he and Dr B. prescribed sugar of lead [lead acetate]. Felt much better after bleeding.' The next day he wrote: 'Much better. Bleeding, sugar of lead, and starvation for ever.' Nevertheless he stayed in bed for six weeks, passing his time writing letters, sketching, studying German and algebra and occasionally receiving visits from miners who related their adventures in the barges.

His personal journal during this period of convalescence recalls his adventures in the tunnel and some of his strengths and weaknesses:

April 22 1828

Here I am in bed at Bridge St. I have now been laid up quite useless for 14 weeks and upwards – ever since 12 January. I shan't forget that day in a hurry. Very near finished my journey then. When the danger is over, it is rather more amusing than otherwise. While it existed I can't say the feeling was at all uncomfortable. If I was to say the contrary, I should be nearer the truth in this instance. While exertions could still be made and hope remained of stopping the ground it was an excitement which has always been a luxury to me. When we were obliged to run, I felt nothing in particular; I was only thinking of the best way of getting us on and the probable state of the arches. When knocked down, I certainly gave myself up, but I took it very much as a matter of course, which I had expected the moment we quitted the frames, for I never expected we should get out. The instant I disengaged myself and got breath again – all dark – I bolted into the other arch. This saved me. By laying hold of the rail rope – the engine *must* have stopped a minute – I stood still nearly a minute. I was anxious for poor Ball and Collins, who I felt too sure had never risen from the fall we all had – and were, as I thought, *crushed* under the great stage. I kept calling them by name to encourage them and make them also (if still able) come thro' the opening. While standing there the effect was – *grand* – the roar of the rushing water in a confined passage and by its velocity rushing past the opening was grand, *very grand*. I cannot compare it to anything – cannon can be nothing to it. At last it came bursting thro' the opening. I was then obliged to be off. But up to that moment, as far as my sensations were concerned, and distinct from the idea of the loss of 6 poor fellows whose death I could not then foresee except this. The sight and the whole affair was well worth the risk and I would willingly pay my share £50 about of the expenses of such a 'spectacle'. Reaching the shaft,

I was much too bothered with my knee and some other thumps to remember much.

If I had been kept under another minute when knocked down I should not have suffered more, and I trust I was tolerably fit to die. If therefore the occurrence itself was rather a gratification than otherwise and the consequences in no way unpleasant I need not attempt to avoid such. My being in bed at present, tho' no doubt arising from the effects of my straining, was *immediately* caused by me returning too soon to a full diet at Brighton: Had I been properly warned of this, I might now have been hard at work at the Tunnel.

But all is for the best. I have formed many plans for my future guidance which I verily believe I shall follow and if so my time will not have been lost. Let's record a few of them they will thus serve to keep me in check as I cannot then deny them. First rules as regards my health I will (if I can) go to bed at such time as to be able to rise early for instance I think I *could* always go to bed at 12 or 1 and get up at 5 or 6 this would agree pretty well with my *shift*-duties and I should then get an appetite for breakfast, which I never used to do. If then I rose early I would breakfast at 8 and eat a substantial one. I could then when I wanted go to town pretty early. Oh that I had a gig or horse!!

I would dine at about 4.30 or 3.30, according to circumstances have tea at 8.30 and a *light* supper at 10.30. This arrangement is peculiarly suited to my occupations, for instance all Dietists seem to agree in one point viz that after a meal you should remain quiet (not asleep however) for one two or three hours. As I have a remarkably good digestion one or one and a half hours would be sufficient. After this period exercise is *necessary*. Now my duties below will be best performed by visiting them at the end of the shift then giving directions for the ensueing [*sic*] shift and going down in the middle to *find fault* &c &c and again at the end as before. The middle visit should *alone* be ever dispensed with. The Duties above ground would require attention more particularly before breakfast and then at any time.

By rising early I can go below to see the end of the nights [*sic*] shift give directions for the next – then attend to above ground matters – &c and at 8 eat a hearty Breakfast.

After breakfast write letters – draw - and attend to any sedentary business. If anything particular go below at 10.30 or so. Then attend again to active business going below again toward 1.30 giving directions for the other shift, then dine at 3 (on these *regular days*). After dinner *write my journal* (standing) – read -&c – draw – attend to office business from Memoranda, always having Morgan Orton &c into me. N.B. this will save me a great deal of time. If nothing particular below – continue these *amusements* till tea at 8, going below again at 9 about. Shift changed &c come into a *comfortable* supper at 10.30. Reading *and enjoying myself* till 11.30 or 12. Go below. Come up and write my journal and Memoranda and go to bed.

Now as I have a habit of eating quick reading at my meals will be a pleasant and excellent thing but having a bad memory I must take notes. Another very good habit I shall thus eat slower which Paris [J.A. Paris *A Treatise on Meals*] says is the way not to eat too much, and moreover enjoy my meals.

Another thing is always to have my journal, a memoranda book and a general sketchbook at hand – (locked up tho;) N.B. In writing my journal I will always open at once an index at the end, and also refer to the page of my sketch book. My memoranda book to enter every thing as it strikes me and then have a pocket one to carry with me. My sketch book will I think be one of Hawkins' letter books. Then by drawing on *whole* foolscap sheets pinn'd on a drawing board I can enter them afterward. All calculations I will enter *arranged* into my – Miscellany. I will have a desk which I *can raise* high enough to stand to. Q: how? – memo – make a sketch. I will write my journal whenever I have an opportunity. I must try and introduce order in everything. I am decidedly improved in that respect within the last two years. As I have a bad memory for words or chronologically a deficiency of language I must make up for it by writing and sketching everything I wish to remember. But I will also try to improve the faculty.

Highly dejected, and envious of his contemporary engineers, he penned:

May 6-7 1828
Here I am again - 12.30 a.m. - Wednesday 7 May. In bed smoking some excellent canaster J. Hulme has left me. He went to Manchester today having spent a day in town on his way from Berlin. Poor fellow he has had some unfortunate *downs* as well as others. It makes me rather blue devilish to think of it and since I am very prone to build airy castles I will now build a few blue ones which I am afraid are likely to prove less airy and more real. Here are these directors damning the Tunnel as fast as they well can. If they go on at this rate, we must certainly stop, and then, by jove we shall stop - payment - where the devil the money is to come from in that case, I know not. Tawny [a director?] is sneaking. We must not expect much from that quarter after a short time. What money we may get from the Battersea concern [the sawmill] will produce at most £300 a year - most likely not £200! The gaz we may never realise, even if we find means to prosecute the experiments difficulties may arise too render the ultimate success doubtful nay perhaps impossible. Where then will be all my fine castles - bubles [sic] well, if it was only for myself I should not mind it. I fear if the Tunnel stops I shall find all those flattering promises of my friends will prove - friendly wishes.

The young Rennies, whatever their real merit, will have built London Bridge, the finest bridge in Europe, and have such a connection with the government as to defy competition. Palmer has built new London Docks and thus without labor have [sic] established a connection which ensures his fortune. While I - shall have been engaged in the Tunnel which failed - which was abandoned - a pretty recommendation, all the engineers glad enough of publishing that it was my father's debut almost in 'engineering'. I have nothing after all so very transcendent as to enable me to rise by my own merit without some such help as the Tunnel. It's a gloomy perspective and yet bad as it is I cannot with all my efforts work myself up to be *down* hearted. Well, it's very fortunate I am so easily pleased. After all let the worst happen – unemployed, untalked of, *pennyless* [sic] (that's damned awkward). I think I may depend upon a home at Benjamin's. My poor father would hardly survive the Tunnel. My mother would follow him. I should be left alone. Here my invention fails, what would follow I cannot guess. A war now, I would go and get my throat cut and yet that would be foolish enough – well '*vogue la galére*', very annoying but so *it is*; I suppose a sort of middle path will be the most likely one. A mediocre success – an engineer sometimes employed, sometimes not – 200 or 300 a year, that uncertain. Well. I shall then have plenty to wish for and that always constituted my hapiness [sic]. May I always be of the same mind and then the less I have the happier I shall be.

I'll turn misanthrope, get a huge Meerschaum, as big as myself and smoke away melancholy. And yet that can't be done without money and that can't be got without working for it. Dear me, what a world this is where starvation itself is an expensive luxury. But damn all croaking. The Tunnel must go on, it shall go on. By the by, why should I not get some situation, surely I have friends enough for that. Qy: get a snug little berth and then a snug little wife with a little somewhat to assist in housekeeping? What an interesting situation!

No luxuries, none of your enjoyments of which I am tolerably fond? – Oh horrible – and all this owing to the dam'd directors who can't swallow when food is put into their mouth. Here is the Duke of Wellington speaking as favourably as possible, offering unasked to take the lead in a public meeting and the devil knows what, and they let it all slip by as if the pig's tail was soaped. Oh for Sir E [Director Sir Edward Owen?] here now, he'd give it to them. But they are all asleep, Hawes and all – all alike – if the tunnel does go on no thanks to them. Well then for a good 4 years yet I'm pin'd down and when finished by that time all the éclat will be gone, all the gilding tarnished, and I shall find as I saw before all my fine castles – gone. – well it's all for the best it may damp a little my vanity – make me a better fellow – and who knows trumps, may turn up again. I'm sure we've no reason to complain yet whenever we've been worse off something good has come.

Regarding his friend Brunel, William Hawes said that deprivation of sleep did not make him irritable and that he was 'joyous, open-hearted and considerate'. Hawes also recalled:

From 1824 to 1832 he joined his friends in every manly sport; and when, after his accident at the Tunnel, he was obliged to withdraw from more violent exercise, he was still ready to co-operate in the arrangements required to give effect to whatever was in hand. To ensure the success of his friends in a rowing match against time, from London to Oxford and back, in 1828, he designed and superintended the building of a four-oared boat, which, in length and proportion of its length to its breadth, far exceeded any boat of the kind which had been seen on the Thames.

The freshness and energy with which he joined in the amusements of his friends after many consecutive days and nights spent in the Tunnel – for frequently he did not go to bed, I might almost say, for weeks together – surprised them all.

Brunel rose from his sickbed on 4 May 1828 and took gentle walks in Blackfriars. As his strength slowly returned he continued experiments with the gas engine and sketched improvements to the triangle-frame engine. On 12 July 1828, he boarded *The Thames* for a voyage to Plymouth. Arriving on 17 July 1828, he spent an interesting ten days inspecting the dockyard and breakwater, also visiting Saltash where he considered the river 'much too wide to be worth having a bridge'. Interestingly, eighteen years later he had different ideas and constructed the Royal Albert Bridge.

For the remainder of 1828 and during most of 1829, Brunel was without a regular occupation but fully engaged in scientific researches and in exchanges with Charles Babbage, Michael Faraday and other friends.

An episode in the bitterly cold winter of 1829 reveals the kindly side of Brunel. One snowy night, his coach overtook another, in the basket of which was an artillery officer, dead drunk, alone and in danger of falling out. Brunel stopped the coach and showed the driver how to tie the man safely and tightly into his seat.

In 1829, Brunel became a member of the Institution of Civil Engineers and with his father attended lectures of the Royal Society; in 1830 he was elected a Fellow and became colleagues with such dignitaries as the previously mentioned Charles Babbage and Michael Faraday.

The idea of a gas engine, which Marc and Brunel had been working on since 1823, was promising and the Admiralty made a grant towards the experiments carried out in 1829 at the abandoned tunnel works at Rotherhithe. The gas was condensed at a pressure of 300 atmospheres and the pipes were required to withstand pressure of 1,500 lbs per square inch – this in an age where steam pressure of 50 lbs per square inch was considered high. The state of contemporary metallurgy was insufficient for the task, and on 30 January 1833 Brunel wrote in his diary:

> Gaz – After a number of experiments I fear we must come to the conclusion that (with carbonic acid at least) no sufficient advantage on the score of economy of fuel can be obtained.
>
> All the time and expense, both enormous, devoted to this thing for nearly 10 years are therefore wasted ... It must therefore die and with it all my fine hopes – crash – gone – well, well, it can't be helped.

Marc and Brunel were unaware that a contemporary scientist, Sadi Carnot, was establishing the laws of thermodynamics which stated that an engine could only produce the amount of energy that was put into it – it could transform energy, but never create it. Had they been aware of this fact, they would not have tried to achieve the impossible.

On 10 June 1830, aged only twenty-four, Brunel was elected a Fellow of the Royal Society in recognition of his work on the Thames Tunnel, his plans for Clifton Bridge and his experiments with the gas engine. He subsequently became a member of most of the other scientific societies, but rarely attended meetings except those of the Institute of Civil Engineers.

In 1830, Brunel joined the Surrey Yeomanry, attended drill and was out with the troop to which he belonged on several occasions.

Marc filled the breach as with the previous one, but investors had lost confidence in the project and funds for further work soon dried up. Stone pillars were erected between the shield's side members and the tunnel brickwork; the ends of the archways were then sealed up. As the visitors' shillings were still most welcome, Brunel set the workmen who still remained the task of decorating the end of the visitors' archway, a large mirror being placed on the new end wall to create the illusion of a continuous arcade. Further gaslights were added and the plant converted to produce gas from the cheaper coal, rather than coconut oil.

On 6 December 1831, Brunel entered in his diary:

[The] Tunnel is now I think dead.
The commissioners have refused on the grounds of want of security – this is the first time I have felt able to cry at least for these 10 years. Some further attempts may be made – but – it will never be finished now in my father's lifetime I fear. However, *nil desperandum* has always been my motto – we may succeed yet. *Perseverantia.*

The Times referred to the tunnel as 'The Great Bore' and the incomplete tunnel became the butt of comedians' jokes. Thomas Hood, humorist and poet, writing in his *Ode to Monsieur Brunel*:

I'll tell thee with thy tunnel what to do;
Light up thy boxes, build a bin or two,
The wine does better than such water trades,
Stock up a sign, the sign of the Bore's Head:

I've drawn it ready for thee in black lead,
And make thy cellar subterranean – thy Shades!

To complete the story, work on the Thames Tunnel lapsed for some seven years, and when work restarted Brunel was busily engaged in other projects. In the spring of 1841, Brunel's three-year-old son Isambard Junior was handed through a gap in-between the two sections of the tunnel – so effectively he became the first person in history to pass beneath the Thames. When the longest underwater tunnel in the world was completed in March 1843 it was for pedestrians only, finance not allowing money to build the ramps at each end to permit vehicular access. Having to use the ninety-nine steps at each end deterred some from using the tunnel and so they continued to patronise the existing ferries. Although the official death toll of workers cutting the tunnel was recorded as seven, Robin Jones in *Isambard Kingdom Brunel* believes that it was more likely to have exceeded twenty, and, when taking into account illnesses caused when cutting the tunnel, the figure could have been close to two hundred.

To boost the number of visitors, the tunnel was turned into the world's first underwater shopping mall and did a good trade selling souvenirs. It was not until 1865 when the tunnel was purchased by the East London Railway that its cost proved justified. The East London Railway ran from the Great Eastern Railway's terminus at Liverpool Street to New Cross, to connect with the London, Brighton & South Coast Railway and also the South Eastern Railway. A track was laid in each of the tunnel's archways and in 1913 the line was electrified.

At the time of writing, funds are being raised to restore the tunnel's splendid entrance hall to its former glory. The public has not used this original grand entrance shaft since the tunnel was adapted for railway use, but it was retained for ventilation purposes and covered during the Second World War to prevent bombs from entering the tunnel.

4

AN ENGINEER IN BRISTOL
1829-31

Ceasing work on the Thames Tunnel had its advantages. Brunel had time to make a trip to France and back and then to pursue other work, such as the gas engine, Bristol Docks and the Clifton Suspension Bridge. This brought him to the attention of those who were to inaugurate the Great Western Railway.

In January 1830, Brunel applied for the post of engineer to the Newcastle & Carlisle Railway which had secured its Act in May 1829. Surprisingly, the job was given to Francis Giles, a very simple canal engineer who had already made himself foolish by declaring that George Stephenson's method of crossing Chat Moss was quite impracticable.

A further disappointment was the design and construction of an observatory at Kensington for Sir James South, a keen amateur astronomer. This included a revolving dome with telescope aperture covered by shutters. Brunel took dimensions on 5 October 1830. All parts were made by Maudslay, Sons & Field. One night, when it was still under construction, a fierce gale sprang up and Brunel feared for its safety. He left Bridge House several times and walked to Blackfriars Bridge to judge the wind's strength, and at midnight decided to walk the four miles to Kensington. As he climbed over the fence to reach the observatory, a patrolling constable observed what seemed to him to be very suspicious behaviour, but accepted Brunel's story and was inveigled into helping secure a tarpaulin.

The observatory was completed on 20 May 1831 and a celebratory opening held, but then that autumn an anonymous article in the *Athenaeum*, probably written by Sir James, criticised both Brunel's design and charges: his estimate had been £504 while in reality it cost in excess of £1,700, and the shutters, estimated at £40, cost over £500. The article described the observatory as 'an absurd project [which] had no other object than the display of tour de force, and was an effort to produce effect on the part of the architect'.

Brunel found this criticism painful and seriously considered taking libel action, but in the end took no further steps. Sir James South initially refused to pay – and monetary disputes became all too frequent in Brunel's life. Brunel felt dejected, for he had little to show for his life so far: an uncompleted Thames Tunnel; a hopeless gas engine; two abortive dock schemes and a dispute regarding an observatory. But he was not a man to stay in the depths of despair; he rapidly recovered from defeat and energetically started on his next task.

In November 1830, Brunel visited Birmingham and carried out surveying for the Bristol & Birmingham Railway. On 10 November 1830, he wrote to his father's friend Charles Babbage:

> You know how important such an opening into the Railway world is to me, and therefore you can conceive how grateful I am towards you to whom I owe it. I trust I shall not disgrace your recommendation.

Unfortunately the scheme proved abortive as the Grand Junction and the London & Birmingham railways were taking up all the available finance. Also in 1830, Brunel was commissioned to carry out drainage work at Tollesbury, Essex. This involved installing a pumping engine and was again constructed by Maudslay, Sons & Field.

Brunel's illness following the Thames Tunnel accident and his consequent ensuing idleness made him depressed, and on 2 August 1832 he wrote to his friend Benjamin Hawes, later MP for

Lambeth: 'Ben I have a painful conviction that I am fast becoming a selfish cold-hearted brute. I'm unhappy, exceedingly so.'

After his recovery from the injuries sustained in the Thames tunnel, it is believed that Brunel's parents sent him to Clifton, Bristol. Back in 1753, William Vick, a Bristol wine merchant, had left £1,000 to the Society of Merchant Venturers; the capital and interest was to accumulate until it was sufficient to build a bridge across Clifton Gorge. By 1829, £8,000 had been accumulated and it was decided that a start would be made, in view of the fact that suspension bridge design had proved that large span bridges could be constructed more cheaply than those with a masonry arch. Land was purchased and designs sought, the additional money required to be raised by subscription.

The concept appealed to Brunel, for recall that as a youth he had helped his father Marc with the drawings of the Ile de Bourbon suspension bridges. For his Clifton project he obtained advice from both his father and the engineer Joshua Field and spent many hours calculating stresses. He spent two days visiting Thomas Telford's Menai Bridge. He also noted that the rhythmic tread of soldiers marching across the Broughton Suspension Bridge near Manchester had caused the pin in one of the suspension chains to snap, thus leading one end of the bridge to collapse.

In November 1829 the Merchant Venturers received twenty-two designs for the Clifton bridge. The judge, Thomas Telford, rejected all and disliked every one of Brunel's four designs. This was because his main spans varied from 870 feet to 916 feet, whereas Telford believed that 600 feet was the maximum for safe lateral wind resistance. Actually, Brunel had taken wind resistance into account and used extremely short suspension rods at the centre of the span, thus bringing the chains down to almost deck level, and by the use of transverse bracing and the addition of vertical chains as used by his father with the Ile de Bourbon bridge.

While staying at Clifton Brunel naturally took the opportunity of visiting Bath, arriving there on 5 July 1830. Making a further visit on 17 September 1830 he called on William Beckford, a wealthy man interested in architecture and engineering and the individual who had built a tower on the hill above his house in Lansdown Crescent, as well as the great Gothic folly of Fonthill Abbey, Wiltshire. No evidence has come to light, but it is quite feasible that Beckford would have been most interested in the Clifton Bridge project and a likely subscriber.

Dr Andrew Swift in *Brunel Comes To Bath* posits the likely theory that when Thomas Telford rejected all the designs for Clifton bridge and was invited to design one himself, instead of producing something aesthetically pleasing like his bridge across the Menai Strait, he produced a hideous Gothic monstrosity, perhaps hoping to impress Beckford. If so, he did not succeed, as Beckford hated it. Telford's design had two enormously tall towers which apart from being expensive to build, would have sunk into the soft ground. As Telford's design was too costly, a new competition was held in October 1830. Respecting the committee's view that a short span would be better, Brunel managed to reduce it to 630 feet, and his design was accepted on 18 March 1831. He used an Egyptian pylon style for his revised plan for the Clifton Bridge, rather than his earlier Gothic designs. On 27 March 1831, Brunel wrote to his friend Benjamin Hawes:

> The Egyptian thing I brought down was quite extravagantly admired by all and unanimously adopted; and I am directed to make such drawings, lithographs, etc. as I, in my supreme judgment, may deem fit; indeed, they were not only very liberal with their money, but inclined to save themselves much trouble by placing very complete reliance on me.

In June 1831, Brunel made a further visit to Bath where he visited the Bath Races and 'met a number of Bristol people – did a little bridge business'. On 11 July he visited Murhill quarry just east of Bath, to inspect the stone possibly to be used for the bridge, and found it 'excellent … very hard and very durable but too white.'

The ceremony of turning the first sod of Clifton Bridge was held on 21 July 1831 when Sir Abraham Elton of Clevedon Court proclaimed of Brunel:

> The time will come when, as that gentleman walks along the streets or as he passes from city to city, the cry will be raised: 'There goes the man who reared that stupendous work, the ornament of Bristol and the wonder of the age.'

Only half the estimated cost of £52,000 for building Clifton Bridge was available, so the project was left in abeyance.

Brunel was incidentally in Clifton at the time of the 29–31 October 1831 Bristol Riots, caused by the monopolistic behaviour of the Bristol Corporation, when a legitimate political protest against the Reform Bill degenerated into a civil riot. He entered in his diary: 'Went to the Mansion House. Found it nearly deserted. It had been broken into and sacked.' Using the back of a broken chair he entered the building and with two others salvaged pictures and plate, carrying them out across the roof and into the Custom House. He managed to arrest one looter but while marching him to a magistrate, another man acting as a constable, but whom Brunel had seen stealing, wrested him away. Brunel managed to secure him again, but yet again he was seized by the pseudo-constable and made a permanent escape.

The following day Colonel Brereton and his troops quelled the riot, but as he had withdrawn them earlier, he was consequently court-martialled for negligence. Brunel was called as one of the witnesses, but Brereton shot himself through the heart before he could be sentenced.

In 1831 Brunel, who shared a love of horses with his father, contributed a forty-five-page article for a volume of *The Horse*, published by the Society for the Diffusion of Useful Knowledge. He analysed its power as a draught animal and gave close attention to the design and construction of wheels and carriages for road transport. In the section on resistance of barges hauled along canals, his thinking on this was applied in due course to his steamships.

One noteworthy aspect about the life of Brunel was the amount of travelling he undertook – and at the drop of a hat. In an era when many did not travel more than ten miles from their birthplace, Brunel travelled all over the country using coaches which charged high fares and were uncomfortable. He first heard of the Monkwearmouth Dock scheme on the north bank of the Wear estuary on 14 November 1831, and, wasting no time, booked a seat for £7 9s 6d on the coach for the evening of the next day. (To put this price into context, a contemporary farmworker earned ten to twelve shillings a week). He had breakfast at Grantham, dinner at York at 4.40 p.m. and arrived at Newcastle at 2.30 a.m. on 17 November, where he had a brief sleep at the Queen's Head before meeting the Dock Surveyor at 8.00 a.m. He had covered 268 miles. He was appointed, but not given any expenses. His plans for a tidal basin and a locked harbour covering a total of 25 acres were rejected by Parliament as being too ambitious, but shortly afterwards a private company was formed for the construction of a dock designed by Brunel. This project was on a much smaller scale than the 1831 plan, and proposed a dock of about 6 acres with a tidal basin of about an acre and a half. Work began in 1834. Unlike his other dock gates, they were in kyanised wood rather than wrought iron. (Kyanisation was a wood-preserving process invented by Dr John Howard Kyan and patented by him in 1832). The dock eventually became the property of the North Eastern Railway.

Brunel could never be criticised for not making the most of his opportunities. While in the north-east he used the occasion to visit Durham Cathedral to study its architecture. When he drove to Newcastle to deposit the dock plans, he also visited the Scotswood Suspension Bridge over the Tyne. When he left Newcastle on 2 December 1831, at Stockton he inspected the very first suspension bridge in England, this carrying the Stockton & Darlington Railway over the Tees. He was unimpressed and noted that two coal wagons depressed the deck twelve inches, commenting: 'Wretched thing, the floor creaks most woefully in returning.' Not long after his visit it required supporting piles and in 1844 the Stephensons replaced it with a cast-iron bridge. He visited Beverley Minster, considering it 'very fine', before calling at Hull to inspect the docks.

On 5 December 1831 he made his first railway journey, when he experienced travel on the Liverpool & Manchester Railway opened the previous year. The entry in his diary read: 'Went by first railway coach to Liverpool 1 hr 25 min on the road and 2 hrs 15 min between Hotels [stations].' A sheet of paper bore a wobbly sketch to which he appended the note: 'Drawn on the L. & M. railway 5.12.31. I record this specimen of the shaking on Manchester railway. The time is not far off when we shall be able to take tea or coffee and write while going noiselessly and smoothly at 45 miles per hour – let me try.' Here we see an instructive example of Brunel's interests broadening from civil to mechanical engineering.

Initially, the Liverpool & Manchester Railway's competition had forced the canal companies to lower their tolls to such an extent that the Liverpool & Manchester Railway could not compete. Brunel commented: 'Went to the Railway establishment, nothing doing apparently in the way of conveyance of goods.' He then visited the Liverpool Docks and the new Custom House, criticising the latter: 'What an extravagant waste of strength in the massive corners and spires – very inferior stone?' Moving on to Chester he saw a work on which his father had advised, the Grosvenor Bridge over the Dee. It was nearly finished, with Brunel commenting: 'A most beautiful, bold and grand work – decidedly the finest and largest in the world.'

On 12 February 1832, Brunel visited William and Elizabeth Horsley at 1 High Row (later 128 Church Street), Kensington Gravel Pits. William was a music teacher, organist and composer whose glees were pronounced by Mendelssohn to be the most perfect musical compositions he knew. William had five children, the eldest of which was Mary, aged nineteen, who, unlike the rest of her family, lacked artistic talent and vivacity. In fact, it was remarked that 'she had nothing to be proud of except her face and her sisters referred to her as "the Duchess of Kensington"'. The younger Fanny was a talented artist and full of fun. John, the eldest son, was to become a Royal Academician; he was a close friend of

Brunel sharing a mutual love of drawing and painting. Sophy was a brilliant pianist while Charles, the youngest, later developed his musical talents under Mendelssohn at Leipzig.

The Horsleys were visited by many famous musicians, such as Bellini, Brahms, Chopin, Mendelssohn and Paganini. In their drawing room Mendelssohn played his latest compositions and they were the first in England to hear his music for *A Midsummer Night's Dream*. On his return to Berlin, Mendelssohn wrote: 'Was it chance that, during the night, somewhere near Boitzenberg, Mary's dear flowers which I was carrying in my buttonhole and which had kept so fresh during the sea voyage, suddenly smelt as sweet as if she were still sitting near me?'

Brunel, who stage-managed an opera written by Sophy Horsley, had the whimsical notion of reversing the obvious allocation of parts; Sophy took on the role of King Death while he acted as the widow. Mendelssohn enjoyed this bit of fun.

Having made his mark in Bristol through designing the suspension bridge, early in 1832 Brunel was invited by the Bristol Docks Company to improve the harbour and so left London for Bristol on 7 February 1832. Situated eight miles from the sea up a tidal and serpentine river, large ships were forsaking Bristol for ports such as Liverpool. In 1804–9, William Jessop had erected a lock at Bristol, thus creating a floating harbour where ships remained at the same level at the quayside and so were unaffected by the state of the tide. In subsequent years a severe problem was caused when this key harbour became silted up.

In his report of 31 August 1832, Brunel suggested removing mudbanks which made parts of the harbour inaccessible; the raising of Netham Weir to divert more water from the Avon through the Harbour; and altering the Rownham Dam at the lower end of the Harbour from an 'overfall' to an 'underfall'. This meant that some silt was scoured out automatically through sluices each time the tide fell, thus leaving less to be removed either manually or mechanically. In 1835 he designed self-acting flaps like Venetian blinds which the flow of water from the Avon forced to close, thus forcing all the water to the floating harbour, whereas tidal water coming upstream would open the flaps, directing the flow to the Avon rather than into the floating harbour. The flaps were not installed until some years later.

Brunel did not criticise Jessop's previous efforts and wrote:

> I do not pretend to suggest anything which shall produce any extraordinary effect. The most that I can recommend is to extend and carry more fully into effect the general system upon which the Docks were originally designed.

In his 1833 report Brunel recommended a 'trunk' of two underwater fences enclosing a trench, which would increase the scouring by the existing culverts. Brunel designed a steam-powered drag boat moved by chains attached to posts on quays. It had a scraper which stirred up mud and deposited the more solid part at the entrance to the trunk; a wooden culvert opened at low water to flush out mud. As the dock company had failed to manually remove the mud on account of the traders who objected to the required two to three annual scourings, banks had built up. Providing a solution, in 1841 Brunel criticised the dock company for not dragging and clearing the sediment, and suggested that Netham Dam be raised, improvements made to Prince Street culvert, a new culvert made at Bathurst Basin and a second drag boat be placed in operation. In 1844 he designed a scraper-dredger, *Bertha,* which remained in service until 1968.

1844 saw work begin on his new entrance lock, the gates of which were of a novel wrought-iron construction so that at high tide they partly floated, making them much easier to move than conventional gates. The lock was crossed by a road bridge of tubular iron design, a precursor of his much larger railway bridges at Chepstow and Saltash. Due to his railway commitments, the lock did not come into regular use until 1849. The Bristol Floating Harbour closed to commercial shipping in 1975 and little dredging is required today.

Brunel's last entry in his personal diary was on 2 August 1832, and he wills it to his friend Ben Hawes, who later became his brother-in-law:

> I always anticipated pleasure and perhaps instruction in reading over this my journal ... Ben I have a painful conviction that I am fast becoming a selfish cold-hearted ambitious brute – only you don't see it ... I'm unhappy – exceedingly so. How convenient the excitement of this election came just in time to conceal it.

Ben Hawes had stood as Liberal candidate for Lambeth and Brunel gave active support to the campaign, though this was one of the few occasions Brunel gave his explicit political support in this way.

5

THE GREAT WESTERN
RAILWAY 1832–36

In the autumn of 1832 four Bristolians met in an office at Temple Back, a site later to be covered by Temple Meads goods depot, to plan a railway from Bristol to London. The outcome was the formation of a committee representing various Bristol organisations which would provide funds for a preliminary survey. The post of engineer was advertised and offered to Brunel then aged twenty-seven and already well-known in the city for his engineering achievements.

The first mention of a Bristol to London railway appears in Brunel's diary on 21 February 1833, when he visited the office of the railway's solicitor, Robert Osborne, and heard of the proposal. The committee had decided that William Brunton and Henry Habberley Price, who had proposed the scheme the previous year, and W. H. Townsend, who had surveyed the Bristol & Gloucestershire Railway, should make a survey; the committee would then adopt the scheme which proposed the lowest estimate. Brunel's friend Roche, who was on the committee, proposed adding Brunel's name to the list of surveyors. That night, Brunel wrote in his diary: 'How will this end? We are undertaking a survey at a sum by which I shall be considerably a loser, but succeeding in being appointed engineer – *nous verons.*'

On 24 February 1833, Brunel wrote to the Birmingham & Gloucester Railway asking for a testimonial:

A survey is about to be made for a line of railway between Bristol and London – I hope to be employed making it. The committee meet to decide the question on Wednesday next. Of four gentlemen who comprise the sub-committee there are two – both of the Society of Friends – to whom I am only known by reputation and not personally – It is possible that Mr Stourage [Sturge] who took an active part in our Glo'ster & Birmingham Railway may be acquainted with them and if Mr Stourge formed a favourable opinion of my activity on that occasion his writing to those gentlemen may be of great use to me – their names are Mr J. Harford of Harford & Davies wine merchants – and Mr Tothill. The letter to be of real use should be by return of post altho' the next day would be better than never – I have the less hesitation in asking as I think your committee own me some consideration – in asking you to exert yourself for me I must trust your kindness.

Also on 24 February 1833, Brunel wrote to his father's solicitor, Mr Sweets, with the aim of him raising support in the City for the Bristol to London Railway:

You may remember some time ago I suggested to you the probability of something being undertaken to forward the skeme [*sic*] of a Bristol to London railway – The Southampton railway party have been constantly agitating the question here and an engineer of the name of Brunton has been making surveys set the thing going – a Committee formed of deputations from the principal public bodies of Bristol has been sitting for some weeks – money is raised for making the necessary survey for coming before the public and going to 'Parliament' – a survey is immediately to be undertaken – Brunton as first in the field has been applied to but I having much more influence here have within the last week succeeded in causing that I am also applied to and on Thursday the point between us is to be decided – the result will be either that we are both employed to make surveys in competition with each other in which case I shall eventually succeed – or as I have still hopes tho' I am rather late in my application I shall be employed alone. In either case it is of importance to form a party in London which would co-operate in promoting my skeme [*sic*]. I have an idea that the entrance of the railway into London may be on the south side – and that at whatever expense it must be brought to the river either at Rotherhithe or Lambeth – in the latter case to Waterloo & Vauxhall bridges and Landowners on that side must be interested and would I should think advocate the measure.

My object of course is to be the first to form such a party & the question only is in what way can this safely and best be done. I think the time has arrived when it is worth turning the matter over in your mind.

I shall be in Town in the middle of the week but only for a few days and should wish to consult you about it.

When interviewed by the Bristol to London committee, Brunel realised they wanted an economically built line, so on 1 March 1833 Brunel informed the penny-pinching GWR directors:

You are holding out a premium to the man who will make you the most flattering promises, and it is quite obvious that he who has the least reputation at stake, or the most to gain by temporary success, and the least to lose by the consequent disappointment, must be the winner in such a race.

Brunel was experienced enough to know that the cheapest was not always the best and told the directors that he would only survey one line from Bristol to London, and that it would be the best, not the cheapest. He envisaged the GWR as principally a passenger-carrying line achieving high speeds in order to reduce journey times.

On 6 March 1833, Brunel caught the night-mail coach to Bristol, riding outside for economy. The next morning he waited in the office of Robert Osborne, solicitor to the railway committee, awaiting the outcome. At 2.00 p.m. the announcement came: he had been appointed engineer. It transpired later that he had won by a majority of only one vote.

Appointed on 7 March 1833, Brunel and his assistant W. H. Townsend agreed to make a preliminary survey for £500. Time was short because of the threat of a rival line, a branch from Basingstoke on the London & Southampton Railway, required the preliminary survey of almost 120 miles to be carried out by the end of May. Wasting no time, on 9 March 1833 Brunel and Townsend set out to make a survey together following the line of the Bristol & Gloucestershire Railway as far as Mangotsfield, Brunel recording:

> Started with Townsend (who was as usual late). Went up the B. & G. line and then across country by Wick Court and over the hill into Lanbridge [*sic* Langridge] Valley and into the London Road by Bath. A most circuitous line and dreadfully hilly country, but I fear the only line which will take in the present railway. Dined at Bath and rode home by the lower road. This latter line along the Avon valley I think, offers greater facilities.

This it most certainly did because Townsend's route involved a climb of 1 in 71 up the Bristol & Gloucestershire Railway to Mangotsfield; then a rise at 1 in 95 to a 1,518-yard-long tunnel under Tog Hill before descending at the same gradient.

Townsend, Brunel's deputy, was a land surveyor and valuer who had carried out a similar task for the Bristol & Gloucestershire Railway. Although Townsend had been quite capable of laying down the nine-mile-long Bristol & Gloucester mineral line, he had little experience of engineering, and lacked the vision required for a main line such as the Great Western Railway. Brunel wrote: 'How the devil I am to get on with him tied to my neck I know not.' Brunel realised that the only route for a railway between Bristol and Bath was along the flat Avon valley.

While in the Bath area, Brunel killed two birds with one stone and carried out both road and railway work. The Black Dog Turnpike Trust was building what is now the A36 along the Limpley Stoke

Valley, just above the Kennet & Avon Canal, and the latter company was afraid that the new road would make the hillside unstable and cause it to collapse into the waterway. He and William Brunton, a Scottish engineer who in 1808 had invented a steam engine with a pair of mechanical feet helping it to climb gradients, were called on to assess the road's stability. Brunel recorded in his diary:

Started at six thirty on horseback to Bath. Had a fall at starting. Arrived at Mr McAdams at a quarter past nine. After waiting a little for Brunton, breakfasted. Brunton came, proceeded with him and Mc & Son on foot to Claverton. The proposed road ran for a short distance parallel to the canal. The side of the hill is a rotten description oolite laying on clay.

Many slips have occurred owing no doubt to the washing of the clay by the rain and considerably assisted by the bad management of the canal. Blackwell, the canal engineer, a bigoted, obstinate, practical man, says the road will make the hill slip – but couldn't tell us why. Cotterell, a Quaker surveyor attending on the part of Vivian the landowner also opposing, could not or would not either say how or why. Merriman, the canal solicitor, appeared to think his advisers rather unsupported by reason. After a useless discussion during which Merriman twice said that Mr B would not of course be convinced and did not come to be convinced (Mr M the last time tried to say this as a joke tho' perfectly true) we parted. Mr Tring, our solicitor, took my card and promised to let me know in good time when they went into committee. Brunton and I then returned to McAdams, wrote our opinions separately and sealed mine and I returned to Bristol.

Time proved that the canal's engineer was correct and Brunel was wrong; the Warminster Road did indeed slip.

Later Brunel surveyed the route between Box and Corsham, before leaving Townsend in charge while he began surveying at the London end, appointing S. Hughes to assist him there. Brunel tended to spend the day in a coach or on horseback and his nights making calculations and plans, only stopping to eat or rest when he had the opportunity.

As Brunel needed to do so much travelling on public coaches, and because hiring horses had their drawbacks, he designed a specially constructed four-horse carriage where he could sleep and write more comfortably. It contained a bed, drawing board and cupboards for his drawing and surveying equipment, food and drink and a box of fifty cigars. The seat was cleverly arranged to extend into a couch so that he could snatch sleep when afforded the opportunity. He spent many weeks in this black-liveried britzska, nicknamed the 'Flying Hearse' by his staff and the contractors. As the railways extended, Brunel had this carriage loaded on to a flat truck and carried as far as possible by train. His days were long. He wrote to J. R. Hammond, in charge of the London Division of the line: 'Between ourselves, it is harder work than I like. I am rarely much under twenty hours a day at it.'

Before any of his surveyors could enter upon a tract of land, Brunel had to visit the landowners and charm them to give permission for a survey; furthermore he had to try and persuade them to take up some of the £2.5 million pounds' worth of shares. It was surprising, in view of his generally impulsive and forthright nature, that he was able to handle landowners with patience, tact and eventual success.

As Brunel was now made engineer to a significant company, it was no longer practical to use his father's office and sleep at Bridge Street or Barge House. He moved to 53 Parliament Street, where he set up a small house and office close to the Houses of Parliament. He appointed Joseph Bennett, aged forty, as chief clerk and hired draughtsmen as required. Joseph remained in this position until Brunel's death.

Unlike his contemporary railway engineers, Brunel insisted on designing the architecture in addition to being responsible for the engineering aspects of a project. Although it gave him much extra work, apart from his own satisfaction he said that it saved his employers money. He always carried with him a sketchbook of sheets of squared paper, and had a sheet of sandpaper pasted to the inside cover in order to keep his pencil sharpened to a fine point. He used this book to sketch rapidly the design of a station building, tunnel portal or bridge; the fact that they were made on squared paper made it easy for his draughtsmen to scale the designs to make large working drawings.

It is interesting to record that Brunel even spent the time and trouble to design such minor details as the lamp posts at Bath station. It was Brunel's policy not to accept any employment as a joint or consulting engineer; he had to be in sole charge. Believing the term 'consulting engineer' to be rather loose, he wrote:

> The term 'Consulting Engineer' is a very vague one and in practice has been too much used to mean a man who for a consideration sells his name but nothing more. Now I never connect myself with an engineering work except as Directing Engineer, who, under the Directors, has the sole responsibility and control of the engineering, and is therefore 'The Engineer'.
>
> In a railway the only works to be constructed are engineering works, and there can really be only one engineer.

In order to reach Reading from east of Bath there were two options: either via Bradford-on-Avon and the Vale of Pewsey, or by way of Swindon and the Vale of White Horse. Brunel recommended the latter as it offered easier gradients and better access to Oxford and Gloucester. He worked tremendously hard, as the committee required the preliminary survey to be completed by May. He spent the daylight hours surveying and his nights writing reports and making calculations. As an example of his intensive activity in 1833 his diary records:

> Wedn. April 17th – Started in the evening by Cooper's coach to the Bear at Reading.
> Thurs. April 18th – Got a hack and rode out to Wo. Hill [Woodley Hill] according to the letter from Hughes. After some search found him – on the wrong track. Directed him as he was so far to push on to the Thames across Early Court and Upper Early and the next morning to begin again at

Chapple Green and go on to Shinfield Green where I would meet him on Saturday morning.

I then rode on to Bagshot Heath and returned by a line going at the back of Easthampstead Park. My horse came down at Mitchell.

Fri. April 19th – Up all night in the intention of going by Cooper's coach but it did not arrive till near five and was then full. Went up by 5 o'clock Reading coach; arrived home, dressed, went to BH [Barge House].

Sat. April 20th – Arrived at Reading late. Went to bed. After breakfast went in search of Hughes. After some trouble found him at 'Black Boy', Shinfield, gave him maps. With him to Theale Road and into Pangbourne. Returned to Reading, went to Theale. Met a Mr Keeps who shewed [*sic*] me the new church. Returned to Reading; Hughes came in the evening. Gave him £5.0.0.

Sun. April 21st – Went to church at the great church – Dr Millman. After church lunched. Started on horseback for Wantage – baited at Blewbury and arrived late at Wantage.

Monday April 22nd – Started at 6 a.m. Examined the ground in the neighbourhood of Wantage – breakfasted at Streatley. Determined on the outer line winding round the undulating ground. Returned to Reading, dined, and went to Theale to meet Hughes. After waiting some time gave it up and returned.

Tuesday April 23rd – After breakfast went in search of Hughes; after some trouble found him at the canal between Shinfield and Calcott Mills, a beautiful place this in hot weather. Gave him the line to Wantage.

Travelling could be dangerous. On 6 May 1833 he noted:

Started by Emerald Coach to Newbury. Arrived there, mounted my horse and rode to Uffington, thence to Shrivenham. Slept there and in the morning proceeded to Swindon. Met Hughes there – found letters from Osborne requiring my immediate return to town. George came; our lines nearly meet, but he has been winding round in a most curious manner. Directed him to point out his Bench Mark to Hughes at Wootton Bassett and then return over his ground to Chippenham following a line I traced for him. Rode to Hungerford; thence to Newbury. Just as I got in sight of the Castle my horse came down – cut his knees and forehead dreadfully – just scratched my knee. I never saw a horse tumble over in such an 'abandonné' style, he dirtied himself even over the withers and croup he rolled over so far. Bled him and left him at the Castle. Returned to town in Bristol Mail.

Brunel realised that it was important for a railway to be as straight and flat as possible. As there was a very gradual rise from London to beyond Swindon, being generally no steeper than 1 in 660, beyond Swindon he believed the fall could best be dealt with by having two rope-worked gradients of 1 in 100, one at Wootton Bassett and the other at Box Tunnel. In the event, locomotives developed to such an extent that by the time the line was opened they could cope with these inclines, heavier trains being offered an assistant engine.

On 30 July 1833, the committee held a public meeting in Bristol to gain support. Brunel stood as one of the speakers and recorded in his diary: 'Got through it very tolerably which I consider a great thing. I hate public meetings – it's playing with a tiger, and all you can hope is that you may not get scratched or worse.'

On 19 August 1833, the title 'Great Western Railway' was adopted. On 27 August 1833, Brunel met Charles Alexander Saunders, appointed secretary to the GWR's London Committee, who became one of his most staunch friends and allies. It was decided to proceed with the detailed survey of the line on 7 September 1833, with G. E. Frere appointed as resident engineer of the Bristol Division.

Other examples of Brunel's hectic schedule and its resulting sleep deprivation are shown by the following entries for 1833:

September 14th – Up at 5. a.m. Joined Place & Williams ranged on to the Island east of Caversham. Breakfasted and mounted.

Called on Mr Hawkes, Surveyor, appointed to be with him at 8 a.m. Rode to meet Hughes; found him in barley stubble west of cottage. Directed him how to proceed and to meet me this evening at the Bear [Reading]. Rode then to Purley Hall. Met Mr Wilder just going in; spoke to him; found him very civil; gave him a prospectus. Rode on to Basildon Farm; left Mr Hopkins' note and my card on Mr Stone. Rode on to Streatley; tried in every way to find a line round instead of crossing the river at Goring; found it impossible. On looking at the country from the high hill south of Streatley however, it was evident that much cutting might be saved by passing SW of Streatley Farm and winding a little more east of Halfpenny Lane.

Returned to Reading, went in search of Mr Stone; found he was gone. Called on Mr Symonds. Hughes came at 7.30. Agreed with him that he was to have £2.2.0 a day and pay his own expenses instead of £35 and charges. Pointed out to him the line he was to follow. Took him with me to Mr Hawks to look at his large plan, Mr H – to furnish him with a copy by tomorrow evening and is to make the survey of the line from Sonning to Streatley inclusive with Book of reference, etc., etc. Came to town by Mail.

September 24th – Arrived at Bath (after an all-night coach journey from London). Met Townsend. Breakfasted and started in his phaeton. Went as far as Keynsham; got out and walked over line. Arrived at the valley at Brislington, found the staffs up – all to double the curve agreed on. Could not make him understand the theory or rationale.

September 25th 1833 – Went over to Bath in a gig – arrived there, called at several places but everybody out.

October 22nd 1833 – Started in a chaise to Saltford where we met Townsend, breakfasted together and walked over the ground from there to the point where we proposed crossing the road – gave him directions upon this part and returned to Bath. Gave Bell directions how to proceed and gave him a letter to Goodridge [Brunel's surveyor at Bath].

Returned to Bristol. Arrived in time for the committee, received positive instructions not even to include Queen's Square in the notice, but to stop at Temple Meads. Suggested to them the possibility of stopping at the Angel at Bath to save crossing the river at this point. Lake gave positive instructions to cross into Lord Manvers' Garden Ground [this overruled Brunel's suggestion of a station in a situation avoiding two river crossings].

Brunel had a boyish sense of humour. On the Bath Road just east of Calne is the Cherhill White Horse. The people of the district were violently opposed to the GWR as it would cause them to lose their coach traffic on the London road. Talking this over one evening, someone suggested turning the horse into a locomotive. Brunel, greatly amused at the idea, immediately sketched the horse from memory, roughly calculated its area and arranged a plan for converting it into an engine.

Ten men were to travel there in two chaises and by moonlight to peg and line out the new shape, then cut away the turf and with it cover the parts of the horse not required for the bold new silhouette. From the chimney was to issue a towering column of steam, and inserted underneath in bold characters the offensive letters 'GWR'. It was never actually his intention to carry out the plan, but he laughed over the sensation it would have created.

As finance was proving a problem, on 18 October 1833 Brunel was ordered to 'discontinue his survey between Reading and Bath, and to confine all further expenses to the survey on the two extremities of the line'.

During the committee stage in Parliament a Dr Dionysius Lardner, a popular scientific writer and Professor of Natural Philosophy and Astronomy at the University of London, provided calculations to prove that if the brakes of a Down train failed at Box Tunnel's east portal it would shoot out of the west end at 120 mph. Unfortunately he had overlooked friction and air resistance, and Brunel was able to provide a more accurate set of figures to prove that the speed would not exceed 56 mph.

The committee was still anxious about the safety of the incline, perhaps thinking that a learned doctor would know more about science than a mere engineer. Ever a practical man, Brunel built a model of the cutting at the western mouth of the tunnel with the rails set at an incline of 1 in 90 – indeed, slightly steeper than the actual incline of 1 in 100. Brunel, grandly producing it, said: 'Now, my Lords, kindly point out whether this dreadful gradient is an ascending or a descending one.'

Brunel had cunningly constructed the model so that the rails appeared to fall towards the tunnel portal, even though in reality they fell away from it. When the committee gave the wrong answer, he took a model wagon, filled it with chalk, placed it on the rails at the mouth of the tunnel then released it so that it rolled down the gradient in the opposite direction to that which the committee had expected and deposited chalk all over the table.

Brunel was cross-examined by the Commons Committee for eleven days, the opposition doing their best to unnerve him, make him contradict himself or demonstrate ignorance. Although Brunel was an excellent witness in parliamentary committees, he always declined the lucrative work of a professional witness and made it a rule never to appear except on behalf of undertakings of which he was the engineer. The most eminent witness in favour of the GWR was George Stephenson, who praised Brunel's scheme, declaring it to be better than all its rivals and commenting that the estimate of costs was realistic. Brunel and George Stephenson's son Robert were close friends, consulting each other, visiting and sharing ideas on the other's projects.

Rail certainly had an advantage. A barge took a day to travel between Bristol and Bath whereas a train would cover the twelve miles in thirty minutes, while between Bath and London there would be no problems caused by freezing waterways in winter, or lack of water in summer. The bill passed by the Commons on 3 March 1834, 182 for and 92 against, was unfortunately defeated in the Lords on 25 July 1834.

The London & Southampton Railway, later to become the London & South Western Railway, sought to extend its territory by promoting a line from Basingstoke to Bath and Bristol. William Brunton was appointed engineer for the proposed branch. At its meeting at the White Hart in Bath on 12 September 1834, Brunel spoke, the *Bath Chronicle* of 18 September reporting:

He began by stating that he came forward in obedience to the call of the meeting, but, not having at all expected to be called on, he felt some difficulty in addressing them. He felt also, that gentlemen of his profession were not the best persons to speak on the merits of a line chosen by themselves, as they might be supposed to have favourable bias towards it. He had been engaged, as they had heard, by the corporate authorities of Bristol, to survey the country between Bristol and London, with a view to the adoption of the best possible line for a railroad communication between those cities. In doing so, his instructions were to select such a line as should embrace the greatest possible extent of public advantage. Both in regard to its levels, and the number of important towns and districts through or near which the railway should be carried. He was wholly unbiased in his choice by any other instructions, and began his survey by re-examining a line which had already been selected by Mr. Brunton, following the direction of the Kennet and Avon Canal. Pending this survey, he employed his assistants in taking the levels of a more northern line, in the directions of the Wilts and Berks Canal, and the result of the two surveys showed so manifest a superiority in favour of the latter that he at once decided on its adoption. Without entering into minute details of the comparative merits of the two lines, he would simply state to the meeting a circumstance of which many present were no doubt aware, that the Kennet and Avon Canal had a great number of locks to attain its summit level, so many as 29 within a distance of two miles: notwithstanding the great elevation this attained, the land is so much higher, that the summit level is of necessity carried through a long tunnel. The Wilts and Berks Canal, on the other hand, has but very few locks indeed, and the summit level is carried without cutting, over a flat country, the valley on the north side of the Marlborough Downs being upwards of 170 feet lower than that on the line proposed from Basing to Bath. This fact would sufficiently shew to the meeting the nature of the difficulties to be met with in the two lines; and he would leave them to draw their own conclusions from it. But it was not on the ground of the superiority of the levels alone that he was induced to adopt the northern line. It appeared also, from a calculation of the traffic on that line, as compared with one in a southern direction, that the former was by far the most considerable. The meeting would be able to judge of the correctness of the statement, when he told that for a distance of 75 miles from London it was in the direct line to Gloucester, would communicate with Stroud and Cirencester, would pass through Windsor, Maidenhead and Reading, within eleven miles of Oxford, and would afford facilities for making a branch of only nine miles in length, through very level country, from Chippenham to Melksham, Bradford and Trowbridge. He would here also state that even the engineers who were called to oppose the Great Western Railway in Parliament (of whom there were a great many), admitted that any railway which would be intended as the main branch to connect the West of England with the metropolis should pass in direction north of the Marlborough Downs.

It is interesting that in his argument Brunel criticises the 502-yard-long Bruce Tunnel at Savernake, but makes no mention of the 3,212-yard-long tunnel his route required at Box – a case of mote and beam!

Brunel had evidently swayed the meeting, which passed the resolution that 'this meeting places the greatest confidence in the exertions of the directors of the Great Western Railway, in endeavouring to obtain the best possible line of communication between the cities of Bristol, Bath and London, and, as those gentlemen are principally connected with this neighbourhood, are identified with its interests, and not influenced by the wants or wishes of any other company, this meeting will not consent to support any rival scheme by which the efforts of the Great Western Railway may be rendered nugatory'. Thus the ball was back in the court of the Great Western Railway.

Just over a month later, another meeting was held at the White Lion, Bath. It was chaired by Charles Wilkins, who owned two textile mills at Twerton just west of Bath, and also most of the village. Brunel had intended bypassing Twerton in order to avoid the expense of having to purchase and demolish many cottages and also divert a road. This diversion would have involved building two bridges across the Avon. Brunel was delighted that Wilkins made no objection to the destruction of property and, when asked where those disposed would live, suggested that some of the railway arches of the viaduct carrying the railway through the village be made as twelve two-roomed cottages, each room having a fireplace. The chimney flue ran horizontally to a chimney designed as a buttress to the outside southern wall of the viaduct. The front room facing the road had a window, but the back room was without. The houses fell out of use as dwellings probably in the latter half of the nineteenth century. It is not known why, but perhaps it was because the horizontal flue prevented smoke from escaping readily; certainly vibration and dampness were not problems. In later years, as the cottages on the opposite side of the road were without a garden for drying laundry, the cottages beneath the railway were rented by the cottagers for a nominal sum, while a coal merchant used two for fuel storage. Today they are in use as industrial storage units or workshops.

Due to the position of the cloth mills, the railway was forced through the garden of Twerton vicarage. The patron, Oriel College, Oxford, received £3,150 in compensation, this sum being used to build a new vicarage designed by Brunel, situated nearer the church on land owned by Charles Wilkins. Unfortunately before the incumbent the Revd Spencer Madan and his family moved into the replacement home, navvies cut through the drains, thereby causing an outbreak of typhoid that resulted in the death of the vicar's son. George Earle Buckle, later an editor of *The Times,* was born at the new vicarage on 10 June 1854.

West of Twerton the line curved along the Avon valley to Saltford. A cutting was needed there which would have destroyed the home of one Major James, who demanded an excessive price for his property. Brunel, not wishing to yield, decided to burrow beneath the house, but quickly realised that if he went under the adjacent building he could retain the line's level without causing a dip. A further advantage was that the tunnel would be shorter. As the owner failed to recognise him when he called, Brunel was able to purchase the house and acre of land for a very reasonable £700.

As the Duke of Wellington had supported Marc's Thames Tunnel, Brunel hoped that similarly he would secure support for the GWR. However, he received this chilly reply:

Walmer Castle
September 25th, 1834

The Duke of Wellington presents his compliments to Mr Brunel.

The Duke is at present out of London and he cannot expect that the gentlemen interested in the Great Western Railroad should follow him into the country. The Duke earnestly recommends these gentlemen not to involve themselves in the expense of such a measure as the one proposed without first fixing definitely upon the whole plan, and ascertaining whether it is practicable for them to carry it into execution consistently with the rights of Property and the joint views of others.

Indeed, considering that there are at present two works of this description in the course of being carried into execution, it might be expedient to delay the execution of a third till experience should have been obtained of the value of the two first. However, although the Duke makes these suggestions he might add that he has no interest whatever in the decision of this question which is not equally felt by the whole community.

Reintroduced into Parliament the following year, the Great Western Railway Bill received Royal Assent on 31 August 1835.

George Burke, KC, whose offices were opposite those of Brunel in Parliament Street, wrote an excellent description of Brunel:

For nearly three years, viz. during the contest for the GWR Bill, I think that seldom a day passed without our meeting. He could enter into the most boyish pranks and fun, without in the least distracting his attention from the matter of business.

I believe that a more joyous nature, combined with the highest intellectual faculties, was never created, and I love to think of him in the character of the ever gay and kind-hearted friend of my early years, rather than in the more serious professional aspect.

We occupied chambers facing each other in Parliament Street. To facilitate our intercourse, it occurred to Brunel to carry a string across Parliament Street from his chambers to mine, to be connected by a bell, by which he could either call me to the window to receive his telegraphic signals, or, more frequently, to wake me in the morning when we had occasion to go to the country together, which, it is needless to observe, was of frequent occurrence; and great was the astonishment of the neighbours at this device, the object of which they were unable to comprehend.

I believe that at this time he scarcely ever went to bed, though I never remember to have seen him tired or out of spirits. He was a very constant smoker, and would take his nap in an armchair, very frequently with a cigar in his mouth; and if we were to start out of town at five or six o'clock in the morning, it was his frequent practice to rouse me out of bed about three, by means of the bell, when I would invariably find him up and dressed and in great glee at the fun of having curtailed my slumbers by two or three hours more than necessary.

No one would have supposed that during the night he had been poring over plans and estimates, and engrossed in serious labours which to most men would have proved destructive of their energies during the following day; but I never saw him otherwise than full of gaiety, an apparently as ready for work as though he had been sleeping through the night.

He had a britska, used on our country excursions, which still live in my remembrance as some of the pleasantest I have ever enjoyed.

I have never known a man who, possessing courage which to many would appear almost like rashness, was less disposed to trust to chance or to throw away any opportunity of attaining his object. In the character of a diplomatist he was as wary and cautious as any man I ever knew.

I frequently accompanied him to the west of England, and into Gloucestershire and South Wales, when public meetings were held in support of the measures in which he was engaged, and I had occasion to observe the enormous popularity which he everywhere enjoyed.

One of these members of the public said of Brunel:

His knowledge of the country surveyed by him was marvellously great. He was rapid in thought, clear in his language, and never said too much or lost his presence of mind. I do not remember ever having enjoyed so great an intellectual treat as listening to Brunel's examination.

Parliamentary bills for new lines usually stipulated the gauge of 4 foot 8½ inches, but the London & Southampton Railway Act made no mention of gauge. Brunel considered 4 foot 8½ inches too narrow and favoured 7 foot ¼ inches which would allow for larger and more powerful locomotives. Bigger and thus more economic rolling stock was also less likely to overturn in the event of a derailment.

Had Brunel raised the gauge question at the committee stage it might well have been turned down, but he wisely kept quiet and it was only after the GWR bill had passed the committee stage, with no opportunity for further objections to be raised, that he pointed out the omission. Claiming the London & Southampton Railway as a precedent, he asked that the gauge clause also be dropped from the GWR bill. His request was granted.

On 15 September 1835, he later revealed to the directors his well-thought-out arguments for building a broad gauge railway:

I beg to submit the following observations upon the subject of the width of the rail as explanatory of the grounds upon which I have recommended to you a deviation from the dimensions adopted in the railway hitherto constructed.

The leading feature which distinguishes railways from common roads is the great diminution of that resistance which arises from the friction of the axle trees and more particularly from obstruction on the road. The latter is almost entirely removed in a well-kept surface of a railway and friction may be considered as the only constant resistance.

The effect of gravity when the load has to ascend any inclination is of course the same whatever the nature of the road and depends only on the rate of inclination.

In the present state of railways and railway carriages the constant resistance which we will call friction amounts generally to about 9 lbs per ton although under favourable circumstances it may be reduced to 8 lbs. Assuming the latter as being the least favourable to the view I propose to take of the necessity of further improvement I will apply to the case of the Great Western Railway.

Upon the GWR from Bristol to Bath and from London to the Oxford branch, a total distance of about 70 miles, including those portions upon which two full thirds of the traffic will take place, there will be no inclination exceeding 4 feet per mile which will cause a resistance of only 1 lb and seven-tenths per ton, calling it an even 2 lbs while friction is taken at 8 lbs it appears that the latter will constitute 80 per cent of the whole resistance. The importance of any improvement upon that which forms so large a proportion is obvious but nevertheless according to present construction of railways a limit has been put to this improvement which limit is already reached or at all events great impediments are thrown in the way of any material diminution of the friction and this serious evil is produced indirectly by the width of the railways.

The resistance from friction is diminished as the proportion of the diameter of the wheel to that of the axle tree is increased there are some causes which in practice slightly influence this result but within the limits of the increase which could be required we may consider that practically the resistance from friction will be diminished exactly in the same ratio that the diameter of the wheel is increased we have therefore the means of materially diminishing this resistance.

The wheels upon railways were originally much smaller than they are now as the speed has been increased and economy in power became more important the diameters have been progressively increased and are now nearly double the size they were but a few years ago even upon the Liverpool and Manchester Railway. I believe they have been increased nearly one half but by the present construction of the carriages they have reached their limit.

The width of the railway being only 4 feet 8 inches between the rails or about 4 feet 6 inches between the wheels the body of the carriage or the platform [of a wagon] on which luggage is placed is of necessity extended over the tops of the wheels and a space must be left for the action of the springs the carriage is raised unnecessarily high while at the same time the size of the wheel is inconveniently limited.

If the centre of gravity of the load could be lowered the motion would be more steady and one cause of wear and tear both in rails and carriages would be diminished.

By simply widening the rails so that the body of the carriage might be kept entirely within the wheels the centre of gravity might be considerably lowered and at the same time the diameter of the wheels be *unlimited* [this plan was never adopted as it was found desirable to use wider carriages overhanging the wheels].

I should propose 6 feet 10 inches to 7 feet as the width of the rails which would I think admit of sufficient width of carriages *for all purposes* I am not by any means prepared at present to recommend any particular size of wheel or even any great increase of the present dimensions. I believe they will be materially increased by my great object which would be in any possible way to *render each part capable of improvement* and to remove what appears as an obstacle to any great progress in such a very important point as the diameter of the wheels upon which the resistance which governs the cost of transport and the speed that may be obtained so materially depends.

The objections which may be urged against these alterations are:

1st The increased width required in cuttings, embankments and tunnel and consequently the increased expenses.

2ndly. A great amount of friction in the curves.

3rdly. The additional weight of the carriages.

4thly. The inconvenience arising from the junction with the London and Birmingham Railway [the initial plan was to share Euston station].

1st. As regards the increase of the earthwork – bridges and tunnel – this would not be so great as would at first sight appear – the increased width of each railway does not affect the width between the rails or on either side as the total widths of the bodies of the carriages remains the same and as the slopes of the cuttings and embankments are the same the total quantity would not necessarily be increased above 1/12th and the cost of the bridges and tunnels would be augmented about in the same ratio and such addition has been provided for in the estimate.

2ndly. The effect of friction upon small curves. The necessary radius of curvature will be increased in the ratio of the width between the wheels, viz: 5 to 7, but the portions of total length which is curved to such a degree as to render this effect sensible [noticeable] is so small (not being above ½ miles of the whole line except immediately at the entrance of the depots) that it is not worth considering where a great advantage is to be gained upon a total distance of 120 miles.

3rdly. The additional weight of carriages. The axle trees alone will be increased and they form but a small part of the total weight of the carriage. The frame will indeed be simplified and I believe this will fully counterbalance the increased length of the axle trees. If the wheels are materially increased in diameter they must of course be stronger and consequently heavier but this weight does not effect the friction at the axle trees and not sensibly the resistance to traction while their increased diameter afford the advantages which are sought for.

4thly. The inconvenience of effecting the junction with the London & Birmingham Railway. This I consider to be the only real obstacle to the adoption of the plan – one additional rail to each railway [i.e. track] must be laid down. I do not foresee any great difficulty in doing this but undoubtedly the London and Birmingham Railway Company may object to it and in that case I see no remedy but the plan must be abandoned it is important that this point should be speedily determined.

The cost of the GWR broad gauge was not much higher than that of the comparable standard gauge London & Birmingham Railway – £56,300 per mile against £53,100.

No time was wasted after the passing of the Act on 31 August 1835 authorising the GWR project. On 3 September 1835, Brunel wrote to Robert Osborne and W. H. Townsend ordering them to have the undergrowth cut down at Brislington, in order that he could set out the exact course of the line and decide where the tunnel shafts should be sunk and to ascertain the nature of the ground through which they were to be driven.

The very first contract for the construction of the GWR was let in November 1835. It was for the Wharncliffe Viaduct across the Brent Valley at Hanwell, and Messrs Grissell & Peto began work in February 1836. The 300-yard-long structure has eight brick arches each with a span of 70 feet.

In December 1835, the directors decided to abandon the idea of sharing the London & Birmingham terminus at Euston and resolved upon building their own at Paddington.

On 26 December 1835, in his office in Parliament Street, Brunel took stock of his life to date:

What a blank in my journal! And during the most eventful part of my life. When last I wrote in this book I was just emerging from obscurity. I had been toiling most unprofitably at numerous things – unprofitably at least at the moment. The Railway certainly was brightening but still very uncertain – what a change. *The Railway* now is in progress. I am their Engineer to the finest work in England – a handsome salary – £2,000 a year – on excellent terms with my Directors and all going smoothly, but what a fight we have had – and how near defeat – and what a ruinous defeat it would have been. It is like looking back on a fearful pass – but we have succeeded. And it's not this alone but everything I have been engaged in has been successful.

Clifton Bridge my first child, my darling, is actually going on – recommenced week last Monday – *Glorious!*

Sunderland Docks too going on well.

Bristol Docks. All Bristol is alive and turned bold and speculative with this Railway – we are to widen the entrances and the Lord knows what.

Merthyr & Cardiff Railway – This too I owe to the G.W.R. I care not however about it.

Cheltenham Railway. Of course this I owe to the Great Western – and I may say to myself do not feel much interested in this. None of the parties are my friends. I hold it only because they can't do without me – it's an awkward line and the estimate's too low. However, it's all in a way of business and it's a proud thing to monopolize all the west as I do. I must keep it as long as I can but I want *tools*.

Bristol & Exeter Railway – another too!

This survey was done in grand style – it's a good line too – and I feel an interest as connected with Bristol to which I really owe much – they have stuck well to me. I think we shall carry this bill – I shall become quite an oracle in Committees of the House. Gravatt served me well in this B. & E. Survey.

Newbury Branch – a little go almost beneath *my* notice now – it will do as a branch.

Suspension Bridge across Thames – I have condescended to be engineer to this – but shan't give myself much trouble about it. If done, however, it all adds to my stock of irons.

I think this forms a pretty list of the real profitable, sound professional jobs – unsought for on my part, that is given to me fairly by the respective parties, all, except MD [Monkwearmouth Dock] resulting from the Clifton Bridge – which I fought hard for and gained only by persevering struggles and some manoeuvres (all fair and honest however). *Voyons.*

I forgot also the *Bristol & Gloster Railway* [*sic*].

Capital:	
70,000	Clifton Bridge
20,000	Bristol docks – to come – Portishead Pier
2,500,000	G.W. Railway – to come – Oxford Branch
750,000	Cheltm Railway
1,250,000	Bristol & Exeter do. Do. – perhaps Plymouth etc.
250,000	Merthyr & Cardiff do. Gloster & S. Wales
150,000	Newbury Branch
50,000	Sunderland Docks
100,000	Thames Suspension Bridge
450,000	Bristol & Gloster Railway
5,590,000	[i.e. total pounds]

A pretty considerable capital likely to pass through my hands – and this at the age of 29 – faith not so young as I always fancy tho' I really can hardly believe it when I think of it.

I am just leaving 53 Parliament St where I may say I have made my fortune or rather the foundation of it and have taken Lord Devon's house, No 18 Duke St – a fine house – I have a fine travelling carriage – I go sometimes with my 4 horses – I have a cab & horse, I have a secretary – in fact I am now somebody. Everything has prospered, everything at this moment is sunshine. I don't like it – it can't last – bad weather must surely come. Let me see the storm in time to gather in my sails.

Mrs B. – I foresee one thing – this time 12 months I shall be a married man. How will that be? Will it make me happier?

The choice of the broad gauge was not announced to the general public until the Report of the Half Yearly meeting in August 1836:

Under these peculiar circumstances and with a view to obtaining the full advantage of the regularity and of the reduction of power effected by this near approach to a level, and also to remedy several serious inconveniences experienced in existing Railways, an increased width of Rails has been recommended by your engineer, and after mature consideration has been determined on by the Directors.

Difficulties and objections were at first supposed by some persons to exist in the construction of Engines for this increased width of Rails, but the Directors have pleasure in stating that several of the most experienced and eminent manufacturers of Locomotive Engines in the North, have undertaken to construct them – and that several Engines are now actually contracted for, adapted to the peculiar dimensions and levels of this Railway calculated for a minimum velocity of 30 miles per hour.

These Engines will be capable of attaining a rate of 35 to 40 miles per hour with the same facility as the speed of 25 to 30 miles is gained by those now constructed for other lines.

Brunel's technical argument of reducing friction and the centre of gravity of rolling stock by using wheels of large diameter, and mounting the coach and wagon bodies within rather than above them, proved inconvenient. The argument for larger and more powerful locomotives boasting higher speeds and greater stability became more persuasive. Although the broad gauge cost an additional £500 per mile to construct, he claimed this investment would be 'amply repaid in the first few years of working'. It was a great pity that his directors acceded so readily, as the cost of transhipment at interchanges between broad and standard gauges proved expensive, as was the cost of the eventual gauge conversion. Brunel lacked the foresight to see that, although there was much to be said in favour of the broad gauge, the cost of converting the existing lines of other companies to broad gauge would have been prohibitive, and that with the developing railway network in Britain, through running over the lines of different companies would be essential. The difference in gauge would render this impossible.

To a certain extent the use of mixed gauge – that is, having a third rail at standard gauge distance between the broad gauge rails – solved the problem, but mixed gauge led to complicated track work; at station platforms the third rail had to be in such a position as to bring a vehicle alongside. Another difficulty was that wagons needed to be balanced on turntables, which meant that the standard gauge track had to be gauntleted between the broad gauge, the three rails thus becoming four.

Brunel disliked facing points on a main line as they could be a cause of derailment. At one period at Bathampton, the Up road to Chippenham was mixed gauge and that to Bradford-on-Avon standard gauge. So that the road was never wrong for an Up broad gauge train, only one movable point blade was provided, the wheels on the other rail running on their flanges for about nine feet. The movable blade was the standard gauge rail situated on the inside of the curve.

Brunel certainly did not believe that all railways should be broad gauge. He advised the directors of the Taff Vale Railway that their chiefly mineral line was unsuited to the broad gauge, speed being unimportant, so the directors suggested five feet as an alternative. In April 1839, he wrote to them saying that the small increase to five feet was not justified and 'most decidedly' recommended the standard gauge for which locomotives and rolling stock was more widely available, and suggested a Bury inside-frame engine would be suitable for the line. He also stressed the importance of being able to link with the probability of a railway being built along the coast and wrote: 'This main line will not be a 5 foot gauge; it will either be a 4 foot 8½ inch or 7 foot – the latter you cannot have and therefore the former offers the only chance.'

Due to the length of the GWR and the problem of road travel in pre-railway days, the line was looked after by two committees, one at Bristol and the other at London, each supervising construction on the relevant half. The Bristol committee was permitted a much greater expenditure on station buildings, architectural decorations and ornamental works than its London counterpart.

Brunel was kept busy dashing to and fro surveying the route between Bristol and London, placating awkward landowners, checking for faulty work by contractors and writing letters and reports. Brunel was not only an engineering genius, but also possessed a powerful personality and a capacity for an immense quantity of hard work. He was not particularly interested in food and drink, his only weakness being cigars which he smoked almost incessantly, though in his youth he had enjoyed a meerschaum pipe with friends.

As mentioned, G. E. Frere was appointed resident engineer to the Bristol end, though Brunel was not keen on delegation and retained a responsibility for detail throughout proceedings – accounting for things such as the proper way to lay bricks and the cheapest wood for posts. He wrote: 'It is an understood thing that all under me are subject to immediate dismissal at my pleasure.' Although Brunel had some brilliant ideas, some were very much less so, and had he listened to others would not have made a number of avoidable mistakes.

In a letter of 8 October 1835 to the GWR directors, Brunel favoured amending plans to have a mile-long inclined plane of 1 in 106 at Wootton Bassett in order to reduce the gradients over the rest of the line. During the few years it took to construct the

Great Western, locomotive technology advanced to such a degree that the inclined planes at Wootton Bassett and Box could be locomotive, rather than rope-worked as originally planned for.

Brunel of the GWR and Robert Stephenson of the London & Birmingham had correspondence regarding the joint use of Euston. Although the GWR Act of 1835 gave legal powers allowing for this, the L&B was reluctant to share the station and a clever escape clause was devised to allow a tenancy agreement for just five years. Had the GWR directors and Brunel been alert, a clause in the Act should have laid down the relationship of the two companies. In due course the GWR had its own terminus at Paddington.

At 1 mile 1,452 yards long, and 800 yards longer than any other railway tunnel built prior to the 1840s, Box Tunnel, burrowing through the southern range of the Cotswold Hills, is the major engineering feature on the Swindon to Bath section of the GWR. The tunnel was cut using eight shafts numbered from east to west, the first and last of which were enlarged into cuttings for the entrances so that only six remained. One was subsequently blocked leaving five remaining today.

Work started in 1836, sinking trial shafts to ascertain the nature of the ground. On 13 June, Brunel reported to his directors:

Five temporary shafts have been sunk on the line of the tunnel to various depths varying from 40 feet to 90 feet to determine the position of the strata of the Oolite through which all of them have been carried – a sixth has been found necessary at the west end – before I can determine with sufficient certainly the exact position of the clay and Fullers' Earth which lies under the Oolite and the proportionate length of the tunnel which it will pass through must govern the relative distance of the permanent shafts – this remaining shaft will be worked day and night and as soon as the required information is obtained which I hope will be in a fortnight, we shall be able to prepare and to let the contracts of these permanent shafts which I propose to do separately from the tunnel in order that the materials through which each portion of the latter is to be carried may be ascertained and worked by the parties most accustomed to the particular description of material and not contracted for blindly on a mere speculation.

In the meantime, work was also progressing at the Bristol end of the line. On 18 June 1836, Brunel reported to the GWR directors:

Contracts Nos. 1, 2 & 3 extending further, canal feeder at Bristol to the Cross Post Gate Turnpike near Bath have been let and the works are now in progress of execution & the whole of this extent will according to the terms of the contracts be completed by the middle of February 1838 by which time the other parts of the line between Bristol & Bath may easily be finished without requiring the same degree of exertions which will be necessary to expedite the work upon these three contracts.

No 1B which includes the greatest portion of tunnelling and generally the heaviest work which occurs between Bristol and Bath & extending from the feeder to the east of a wood on the bank of the river commonly called Dr Fox's Wood, a distance of about two miles and three-quarters has been contracted for by Mr Ranger, the notice to commence works was given on 21st March last and from that day the works have proceeded with activity – the time allowed in the contract for the entire completion of the works will require 21st January 1838. The commencement of this contract involves much experience and laborious work without producing a corresponding apparent result, a great quantity of the best material derived from the excavation have to be carried through the tunnels to the works at the other extremity of the line, and consequently the forming a heading or driftway through the line of these tunnels is the first and most important operation – and the period for opening these headings was limited on the contract under a heavy penalty ... In the Tunnel No 1 (the first from Bristol) two temporary shafts have been sunk so that the headings can be carried on at six places at once – one from each end and two from each shaft. In Tunnel No 2, one shaft has been sunk which gives four faces for the heading. In tunnel No 3, which is the longest between Bath and Bristol, being upwards of half a mile, three permanent shafts have been sunk and two temporary ones, so that the headings can be worked at twelve different faces. At eight of these they are in active operation and the remaining four will be very shortly, the two shafts from which they will be worked being within a few feet of the required depth. At these headings the work has been carried on day and night, but I do not expect that they will be completed within the time fixed by the contract which for the tunnels Nos 1 & 2 expires on the 21st of next month. Time has been lost in this and in other parts of the work by the injudicious arrangements of the contractor's foreman. At my request he has been dismissed and the work has since proceeded more expeditiously and with greater advantage to the contractor himself as well as to the company.

As I before observed, until these headings are completed, the progress of the work will not be very apparent. The cutting and the embankment at the west end of Tunnel No 1 is however proceeding with about 200 feet of the embankment being formed. A five foot culvert in this embankment has been finished and the foundations of an accommodation bridge in the meadows are in progress, and the bridge itself will be completed as soon as a larger supply of stone can be brought from the quarry.

Preparations are making for commencing the bridge across the Avon. The excavation for the foundations is proceeding with, and they have begun driving the piles of the coffer dam. The contractor has obtained leave to form a temporary road along the banks of the river from above Netham Dam to the site of this bridge and has laid temporary rails for the carriage of the stone which is quarried in Dr Fox's Wood brought by the river to Netham Dam.

In the Nightingale Valley the contractor has established his principal workshop for the repair of wagons and tools.

In Conham Wood or Brickwood between Tunnels 2 & 3 nothing material can be done until the headings are complete.

In Dr Fox's Wood the embankment is proceeding, the excavation at each end consists of very hard sandstone. From this part it was always expected that we should obtain excellent materials for the construction of the Avon Bridge and the masonry generally, and we have no cause to be disappointed. At the west end particularly a quarry of very fine stone has been opened and is now working and an ample supply of good stone for smaller work may be selected from the excavation at the eastern end.

Upon the whole I consider that the works in No 1B are now proceeding satisfactorily. The contractor I think appears desirous of doing his work properly and with the exception of the inefficiency of the superintendent who he has discharged I have no cause of complaint.

No 2 is contracted for by the same party as No 1 and the notice for commencement was dated 18th May and the whole is to be completed by 18th January 1838. The contract extends from the termination of No 1B to a short distance beyond the Keynsham Brass Mills – a length of about two miles and three furlongs. The principal works upon it are a deep cutting through the hill upon which Lodge Farm is situated and the embankment across the Keynsham Hams – the principal piece of masonry will be the bridge across the Chew and probably a short tunnel through a portion of the cutting which is expected to be of very loose material. The cuttings and embankments are commenced at each point upon the line. The drains and temporary fencing are completed across the Keynsham Meads, the foundations for the Chew Bridge are being excavated and the work generally is proceeding satisfactorily, but there is nothing particularly worthy of remark. No 3B, which extends from the termination of 2B to the turnpike road near Cross Post Gate, a distance of about three miles, has been contracted for by Mr McIntosh. Some delay has taken place in commencing this contract in consequence of the refusal of Colonel Gore Langton or rather his agent Mr Brown to consent to a deviation from the parliamentary plan which I found it would be desirable to make and which I had reported he could not object to.

Another small deviation which does not affect the line at the point where the works must be commenced has been since assented to by Colonel Langton.

The notice to the contractor dates from Wednesday next the 15th and the period for completion will expire February 1838. The principal works upon this line consist of a long cutting through limestone, in the middle of which cutting occurs a short tunnel or covered way under the road at Saltford, and a rather heavy embankment across the Corston and Newton meadows.

An example of Brunel being highly sensitive to the environment was demonstrated during the construction of Bristol No. 2 (or St Anne's Park Tunnel). During construction a landslip carried away a portion of the crenellation of the west portal. It was on the point of being repaired when Brunel said: 'By no means, leave it as it is, train ivy over it and it will appear as a beautiful ruin!' It is now Grade II listed. E. Churton, in *The Rail Road Book of England,* published in 1851, remarked that this tunnel mouth was 'so pleasing an object that it has long been considered one of the principal attractions of the neighbourhood'. The use of 'long' is curious as it had only been in existence for about twelve years.

Brunel was influenced by a book by John Claudius Loudon, *Encyclopaedia of Cottage, Farm, and Village Furniture,* published in 1833. The castellated viaduct west of Bath station is very similar to Loudon's sketch of a villa, while the entrance and departure gateways to Bristol Temple Meads station are almost identical to that same villa.

Brunel's brother-in-law artist John Horsley wrote regarding Brunel:

Being naturally imbued with artistic taste and perception of a very high order ... [he] had a remarkably accurate eye for proportion, as well as taste for form. This is evinced in every line to be found in his sketchbooks, and in all the architectural features of his various works. So small an incident as the choice of colour in the original carriages of the Great Western Railway, and any decorative work called for on the line, gave public evidence of his taste.

Brunel's principal working method was to produce accurate sketches from which one of his assistants would then produce working drawings – and woe betide them if they introduced any inaccuracy!

In June 1836, Brunel interviewed several men who had applied for the position of assistant engineer to the Box Tunnel project, afterwards writing to William Glennie offering him the position. Glennie replied accepting the post of *resident* engineer, so Brunel wrote:

> You have entirely misunderstood me. I think I explained that this position of resident engineer was attainable by you – not promised by me. All I can offer you is the post of assistant engineer. What I offer must not be a certain or permanent position. My responsibility is too great to allow my retaining anyone who may appear to me to be inefficient. It is an understood thing that all under me are subject to immediate dismissal at my pleasure. It is for you to decide if you are likely to proceed satisfactorily and whether the chance is sufficient inducement.

Glennie accepted the post and was later appointed resident engineer at a salary of £150 per annum. His most important task was keeping the tunnel centre line heading in the correct direction and on a gradient of 1 in 100.

In November 1836, under the supervision of Charles Richardson, later responsible for the Severn tunnel contract, eight shafts of Box Tunnel, each 28 feet in diameter and varying from 70 to 300 feet in depth, underwent construction. The shafts were completed by the autumn of 1837 and the tunnel project contracts advertised. Following the plan outlined in the last sentence of his report to the directors of 18 June 1836, George Burge of Herne Bay, then building St Katherine's Dock in London, won the contract for cutting the three-quarters of the tunnel through clay, blue marl and inferior oolite, while the remaining half-mile was cut by Brewer of Box and Lewis of Bath who had the necessary skill for quarrying with great oolite and had already sunk the trial shafts for this portion.

Burge, cutting through ground which needed supporting, probably used the 'English method' of tunnelling perfected by canal builders fifty years earlier. First a pilot heading was driven along what became the tunnel arch. Crown bars were inserted, supported at one end by the brick lining and at the other on timber props. Miners would then widen and deepen the excavation supported by timber. Larch bars were preferred as they gave an audible creak if the load was excessive and collapse imminent; this could then be forestalled with extra props as necessary. Behind the excavators came the bricklayers who added a lining. At no point were excavations left unlined for more than about eight feet ahead of the bricklayers. The workmen lodged in neighbouring villages and, because the work went on both night and day, no bed was ever cold – in fact they played appropriately 'Box & Cox'.

A ton of gunpowder and a ton of candles, the latter made at Box, were consumed on site week. Bricks were supplied by Hunt's brickyard, employing a hundred men in the meadows west of Chippenham. Over three years, 100 horses and carts carried a total of 30 million bricks to the site.

Brunel had learned to appreciate the cost of materials when working on the Thames Tunnel. It happened not infrequently that it was desirable to accept the tender of some contractor for railway work whose initial prices upon an article were too high, with Brunel then going into the details to convince the contractor of his error. He would go step by step through the stages of the work, to the surprise and alarm of the practical man as he found himself corrected in his own special business.

He proved to a Chippenham brickmaker that bricks could be made cheaper than he supposed. Brunel knew precisely how much coal would burn so many bricks; what it would cost; what would be the cost of housing the men; the cost of cartage and how many men would be required to complete the work in the specified time. The contractor finally accepted the contract and made money from it. Similarly with the Maidenhead bridge, Brunel pointed out that the weight which the contractor feared would crush the bricks would in fact be less than in a wall which the contractor had previously built.

Periodically Messrs Lewis & Brewer found themselves being criticised by Brunel, such as on the occasion of 23 March 1839:

Mr Glennie has just informed me that the works have again been delayed by the breaking of your machinery.

This has been of such frequent occurrence and the consequences are likely to be so serious that I must insist upon some decisive steps being taken to remedy the evil. It is not my business to point out to you where is the principal defect. Whether it be in the designing of the machinery or in the construction, or in the use of it, it is evident that it cannot be depended upon. I never knew a work in which the failure of machinery was so frequent – when the Directors desirous of affording you every encouragement agreed to postpone for the present leaving the 1/5th of the amount of penalties due by you it was upon my representation that the works were now in such a state that great progress might be made and that I had hopes that you would soon regain the arrears which had accumulated. It is evident that no such hope can be entertained while the works are subject to be stopped weekly and almost daily by accidents to the machinery and I feel that the Directors will be compelled to enforce the penalties immediately and that I must withhold any certificates until I can report that the machinery upon which everything depends is put into an efficient state of repair – your most serious attention must therefore be immediately given to this subject.

The *Wiltshire Independent* of 11 July 1839 reported, regarding the union between Lewis and Brewer's tunnel workings:

But, on breaking through the last intervening portion of rock, the accuracy of the headings was proved, and to the joy of the workmen, who took a lively interest in the result and to the triumph of Messrs. Lewis and Brewer's scientific worming, it was found that the junction was perfect to A HAIR AS TO THE LEVEL, the entire roof forming an unvarying line; while laterally, the utmost deviation from a straight line was only ONE INCH AND A QUARTER!

This was an immense achievement for a length of 1,520 feet begun at opposite ends. Brunel was so delighted that he took a ring from his finger and presented it to the ganger beside him.

The construction of the tunnel had involved many vicissitudes, moments of mirth and elation as well as taxing struggle. F. S. Williams's *Our Iron Roads* of 1883 relates a charming anecdote:

On one occasion some of the directors of the Great Western Railway were inspecting the works at the Box tunnel, and several of them resolved to descend a shaft with Mr Brunel and one or two of the other engineers, who mentioned the incident to the writer. Accordingly all but one ensconced themselves in the tub provided for the purpose – he declined to accompany them. His friends rallied him for his want of courage, and one slyly suggested, 'Did your wife forbid you when you started?' A quiet nod in response intimated that the right nail had been struck, and the revelation was received with a merry laugh. But as the pilgrims found themselves slipping about a greasy, muddy tub, jolting and shaking as the horses stopped – by whose aid they were lowered – and how at length they were suspended some hundred and fifty feet from the bottom, till the blastings that had been prepared roared and reverberated through the 'long-drawn caverns', more than one of the party who had laughed before, wished that they had received a similar prohibition to that of their friend above, and that they had manifested an equal amount of marital docility.

In another account from the same book, the tone is markedly different:

For a considerable distance the tunnel passes through freestone rock, from the fissures of which water flowed so freely that, in November, 1837, the steam-engine used to pump it out proved insufficient, one division of the tunnel was filled, the water rose fifty-six feet high in the shaft, and it was found necessary to suspend operations till the following midsummer, when a second engine of fifty-horse power was brought to the assistance of its brother leviathan, and the works were cleared. Another irruption took place, and the water was then pumped out at the rate of thirty-two thousand hogsheads a day.

Brunel wrote to Burge regarding this episode on 2 April 1840, commenting on the pumping engines at work in the flooded portion of the tunnel:

> Sir,
> While you are wasting so much valuable time at Shaft no 6 in a bungling attempt to make the present machinery do that for which it is totally unfit, you are neglecting to drive the heading between Shafts 5 and 6 which would have shown some intention of providing against similar difficulties next winter. While such a lack of management continues I shall not recommend to the directors any further payment.

In *A Brief Account of the Making and Working of the Great Box Tunnel* by Thomas Gale, published in 1884, it is stated that 100 men lost their lives during its construction. Gale was a sub-contractor under Burge, but it is believed that in this he made an overestimation, for the GWR reported only ten deaths and the local papers recorded only a handful of fatalities. David Brooke in the *Journal of the Railway & Canal Historical Society* (No 142, July 1989) states that there were only nineteen fatalities while building the tunnel.

Although Brunel for the last twenty-six years of his life had enough railway work to occupy a normal person's career, he was additionally involved in shipbuilding. It all started through a joke. At a GWR directors' meeting in October 1835, when someone objected to the length of the line from London to Bristol, Brunel with characteristic grandeur in mind and with tongue firmly in cheek remarked: 'Why not make it longer and have a steamboat to go from Bristol to New York?'

6

MARRIAGE TO MARY
1836

On 14 April 1836, Brunel reprised his diary entry of 26 December 1835:

> Since that time [December 1835] I have added to my stock in trade the Plymouth Railway, the Oxford branch and today somewhat against my will the Worcester & Oxford. Here's another £2,500,000 of capital – I may say £8,000,000 and really all very likely to go on. And what is satisfactory all reflecting credit upon me and most of them almost forced upon me ... Really my business is something extraordinary.

The last few lines suggest that his financial state could now allow him the freedom to marry. On May 1836, on a walk along Holland Lane, he proposed to Mary Horsley. Her younger sister, Fanny, wrote a very comprehensive letter to her aunt giving a full account:

> It literally came on us all like a thunderbolt, though certainly one of a very pleasant description. I think he called once in March, and that was all till last Thursday week, when he called on his way to Hanwell, and said he would come back to tea at nine o'clock; which he did, and staied [sic] chatting very

pleasantly till eleven. A long time ago he told us he was very fond of musk plants, and Sophy and I had often given him bunches of it out of our garden, and Mary promised that some day or other she would give him a pot of it. So that morning, she got one, and when he was going away she gave it to him, but he said he would rather leave it as he was going to walk home, and would send for it the next day. No one came, however, either Friday or Saturday, and I, as you may imagine, made many wise reflections on forgetfulness and so forth, when on Sunday afternoon he arrived in person. It was near dinner-time and Mamma asked him to stay, which he agreed with great alacrity. At seven o'clock Mr Klingemann and Dr Rosen came to tea, and Isambard expressed a great wish to see Lord Holland's Lanes, so, by way of doing a very genteel thing, we all agreed to go. Isambard offered Mary his arm – Mamma went with John, I with Mr Klingemann and Sophy with Dr Rosen. We walked all through the lanes to the house, and then back. Mr Brunel and Mary walked all the way very slowly, but when, on our return, we were quite at the bottom of Bedford Place, they were only just visible, and Mamma got quite vexed and annoyed, never thinking of the real reason. They were some minutes after us in finding their way up to the drawing room, and when they did enter, Mamma said, 'Upon my word, Mr Brunel, I never knew anyone walk so slowly in my life.'

'Why indeed,' he said, 'I walk so seldom that when I do, I like to make the best of my time,' which as we have discovered, was rather a witty answer. He almost immediately took leave with his musk, and I certainly saw a look which ought to have flashed conviction on my mind, but it did not. Mary was silent and pale all the evening, but I thought nothing of it; Mr Klingemann staid late, and directly he had gone, Sophy and I sent up to bed. In about half an hour we heard Mary come up and called her in.

'Well, what could you be doing lagging behind in that way?' said Sophy.

'Indeed, Mary,' I said – but quite in fun, without any idea of the truth – 'one would think he had been making you an offer.'

'And what would you say if he really had?' said Mary in an awful, hollow voice which I shall never forget. Sophy and I immediately fell into such tremendous fits of laughing, that Mary said she must go away. It certainly seemed very unfeeling – and she, poor thing, with tears in her eyes – but so it was, and I must confess I was much the worst of the two. However, we soon got composed, and listened with delight to the little she had to tell, I mean little in quantity for such happy and excellent facts are great, if anything is. He made her the offer as they were coming home, and told her he had liked her all the five years he had known her, but would never engage her till he was fully able to keep a wife in comfort – I do admire his conduct very much, so honourable and forebearing – not shackling her with an endless engagement, as so many men would have done, but leaving her free, with her mind clear to enjoy pleasure, and to gain improvement and experience during the years of her youth. I do not mean that her youth is over, poor thing, there everything is as it should be, she two and twenty and he thirty. I always thought he admired and paid her more respect that anyone else, but never dreamt of its coming to this. I often said and thought that Mary would have chosen him before anyone else in the world. Well, I hope we are all, as you say in your little note, not only joyful but thankful. Indeed we have a very great cause. He came every day till Friday, and on Friday, melancholy to relate, he was obliged to go to the country, and does not return till Sunday. I think Mary has borne it very well with the constant aid of pen, ink and the post, and sundry double letters. You know I must have a little laugh, but really, *Love* is such a very new character in our family, not to speak of *Marriage*, that I only wonder at my good behaviour on the occasion.

Now I must tell you how delightful it is that all his family approve it so much, and are very kind. Nothing can equal Mr and Mrs Brunel's and Emma's kindness, and Mr and Mrs Benjamin Hawes are just the same. We all had our fears about this till their letters came, which were everything that could be desired. On Monday, Mamma and I and Mary set off at half-past ten to town in a fly. I took a book and well it was I did, for the hours they spent at Turner's making endless substantial purchases would have been unbearable. We got about four to Rotherhithe and found the family at home. What a perfect old man Mr Brunel is! I leave Mary to adore the son, but I really must be allowed to adore the father. Mamma and Mrs Brunel retired to a private conference after some time, and then Miss Brunel and Mary, so he proposed taking me to see the Tunnel, which is only six or seven yards from their house. It was the first time I had seen it, and I cannot tell you how much I was impressed with wonder and admiration at it and at the mind of the man who could conceive it. There are numbers of men at work at it now, and it is going on most briskly, I believe it will be finished in about two years.

Now I think I have told you the *How, When* and *Where, How*, perfectly delightful. *Where*, Sunday week is Lord Holland's Lane. No day is, or indeed can be, fixed, for I suppose it must depend entirely upon his engagements, which, as you know, are numberless and imperative. She is going to be dressed completely in white, and married in Kensington Church, and have the bells rung in the good old style; and there is to be a breakfast, only a quiet family one which is much the best, and then to my particular delight, a gay dance in the evening. They mean to go to N. Wales, and come back through Devonshire. Really everything is charming.

Brunel was not one to waste time. The marriage took place in Kensington church on 5 July 1836. The fortnight's honeymoon by stagecoach started in Capel Curig and ended in the West Country before settling into 18 Duke Street, Westminster, convenient to Parliament and a home they enjoyed for the whole of their married life.

The previous owner of this house in Duke Street was the Earl of Devon. Of an eighteenth-century design in red brick, it was four storeys high. The door from Duke Street was on the upper ground floor, while the lower ground floor was at the level of St James's Park. The drawing room on the first floor was decorated with paintings by Mary's family: two by brother John and seventeen by her great uncle Augustus Calcott.

In December 1835, Brunel moved his office from Parliament Street to Duke Street. A speaking tube ran from Brunel's desk to his assistants in other offices; Brunel's offices were on the lower floors. Needing more office space, he added a detached building 40 feet by 20 feet for draughtsmen and clerks. His own personal office had only one chair – a clever device to prevent time-wasters staying.

The first important event after Brunel's marriage occurred on 27 August 1836, when they travelled to Bristol for the laying of the foundation stone of the Clifton Suspension Bridge on the side of the Leigh Woods abutment. In order to transport men and material between the two abutments, a wrought-iron bar 1,000 feet long and 1½ in diameter was slung between them. It was welded to length in Leigh Woods and on 23 August 1836 hauled across by cable. Unfortunately when it was almost across the span the cable snapped and the bar fell, luckily without causing anyone injury. When raised into position the next day it had a bend in the centre. When most of the spectators had left the site on 27 August, a fellow foolishly entered the basket on the bar and it ran down to the centre where it consequently stopped. As the hauling rope had not been wound in as the basket moved, the rope dangled towards the river. The SS *Killarney* was passing and her mast lassoed the rope. Fortunately someone had the presence of mind to sever the rope, leaving the basket 'swinging to and fro with fearful rapidity', as reported in the *Bristol Mirror*. It was only with 'utmost effort that the gentleman could keep himself in; when it ceased to swing the gentleman was drawn to the rock'.

A new bar was substituted in September and on the twenty-seventh of that month Brunel himself was hauled across at 6.00 p.m. accompanied by the son of his friend Captain Claxton. Brunel made two further trips that evening accompanied firstly by Mr Coulson and then by Mr Tate.

Although two piers were completed in 1840 the rest of the work remained in abeyance during the remainder of Brunel's lifetime. In 1854, the Bridge Trustees collected £125 in fares from those who gained excitement from crossing the gorge in this basket. The bridge was formally opened on 8 December 1864 using the double chains from Brunel's Hungerford Bridge, built between 1841 and 1845, which had been dismantled to make space for Charing Cross railway bridge. The original Bristol chains had been sold for use in the Royal Albert Bridge across the Tamar. Brunel never abandoned hope that the bridge at Clifton would be completed and on 19 October 1849 wrote to his assistant:

My Dear Ward,
Pray do not let anything be done if you can possibly prevent it which can in any degree be a step towards the abandonment of the bridge – out of evil sometimes comes good, and the very badness of the times in all mechanical manufacturing engineering (Engineering properly speaking is not *bad* but *dead*) will be the cause of our finishing the bridge. If I can limit the capital required to manufacturer's capital that is materials and work I can get parties to undertake it, and I think I can manage all (with some little personal sacrifice) except the Ferry.

If you will get that burden or bug-bear – for it is more alarming that it ever ought to be – removed I think I shall see my way set to work now manfully, and you will succeed and then I shall be bound to also.

It is difficult to learn much about Brunel's marriage. He certainly loved Mary, but due to his tremendously heavy work commitments was frequently separated from her. She ran the household very efficiently for him and was remarked upon as always looking elegant. As related by C. Noble's *The Brunels: Father & Son*, 'She had, for morning use only, a carriage lined with green moiré silk while for evening she had another lined with the same in cream colour,' and, 'She never strolled in the park below her windows without a footman in livery following behind.'

Mary bore him three children: Isambard (1837–1902), Henry Marc (1842–1903) and Florence Mary (*c.* 1847–76). The eldest was not interested in engineering and entered the legal profession. Specialising in ecclesiastical law, eventually he became Chancellor of the Diocese of Ely. Henry, on the other hand, was very interested in his father's work. Henry served an apprenticeship with William Armstrong and later became a partner of Sir John Wolfe Barry, who designed the Tower Bridge and the Connell Ferry Bridge in Argyll. Neither son had an issue, but Florence married Arthur James, a master at Eton, and had a daughter Celia. Brunel was fond of children, both his own and his nephew Benjamin Hawes.

Planning the line immediately east of Bath, Brunel met with a problem in the shape of the Kennet & Avon Canal and a steep slope reducing the land available. A public pleasure ground, Sydney Gardens, was furthermore in the way of the railway's route. With great tact Brunel persuaded the canal company that if he diverted the canal it would obviate a sharp curve and thus ease navigation, while he induced the garden trustees that the railway's architecture would enhance their ground. On 23 January 1839, he wrote detailing the proposal to Henry Godwin, secretary to the garden trustees:

I now forward to you according to your request a memorandum of the proposal made by me this morning on the part of the Great Western Railway Company of a modification of the Terms of Agreement existing between the Proprietors of the Sydney Gardens and the Company.

It is proposed that the railway should be carried through the Gardens by an open cutting instead of a Covered Way. The ground on the upper or east side being secured by a retaining wall and on the lower side to be sloped down towards the railway with a terrace & wall at the bottom.

The slopes to be trimmed and soiled and returned to the Proprietors of the Gardens. A bridge to be constructed for the present centre walk of not less than 28ft in width between the parapets. The surface of the walk over the bridge to be made level with the present surface at the east end of the site of the proposed bridge with so many steps at the west end as may be necessary to connect the surface of the bridge with the level of the Walk. Two smaller bridges of cast iron or wood to be constructed at such points as the Proprietors may determine.

So much of the top soil not required for the slopes or before described and one thousand cube yards of any gravel obtained from the excavation to be laid aside and deposited on any part continuous to the line of railway which may be selected by the Proprietors.

The bridge and retaining walls to be constructed according to the general character of the designs submitted this day to the Committee of Proprietors.

Thus Brunel took the utmost care to preserve the gardens' beauty. In fact, James Tunstall in *Rambles in Bath & District*, published in 1847, wrote that the railway and canal 'so far from detracting from, are made to increase the beauty of the promenades'. The high quality of Brunel's works is reflected in the many Grade II structures in the area: the graceful elliptical-arched Beckford Road Bridge, and Sydney Road bridge; the finely dressed retaining wall with its pronounced, curved batter; the unusual balustraded stone wall on the other side of the line and two ornamental bridges. Brunel succeeded brilliantly in running the railway through the gardens without impairing their beauty, while passing trains added to the overall interest of a visit.

In the first week of September 1839, Brunel recorded:

Operations commenced on that part of the line facing Hampton Row, on the bank is of the canal, and many workmen have been employed in digging out the earth from the opposite field, preparatory to turning the course of the canal. The works in the neighbourhood of Wells Road, as regards the great archway and the coffer dam and likewise proceeding most rapidly.

Just as Brunel was to experience initial trouble with his planned way, his locomotives likewise caused problems.

When the GWR ordered its first engines, a locomotive superintendent had yet to be appointed, so responsibility fell to Brunel. Unfortunately his expertise in the locomotive field proved to be inadequate.

He set out his locomotive specifications in an undated letter of 1836:

I am authorised on the part of the Great Western Railway company to apply to you amongst several other Manufacturers to know whether you will undertake & generally upon what terms but more particularly within what period to supply one or two Locomotive Engines. The particular form and construction of engines will be left to your own judgement, the object of the Company being to induce Manufacturers to turn their attention to the improvement of Locomotive engines and to afford them an opportunity of introducing such improvements as may suggest themselves when unchecked by detailed and particular specification of the parts, it being intended that the result of the trials of the various engines thus finished should lead to more extended orders to those manufacturers who shall have made the engines best adapted to the objects of the Company. These objects can be particularly defined but most principally related to the speed – to the related economy of the first construction – of the subsequent repairs and of the consumption of fuel – generally to the performance of the greatest quantity of work at the greatest speed at the least expense and in the most convenient and advantageous manner.

The comparative importance of those objects and the best means of obtaining them will be left for you to determine, but the following are a few conditions which *must* be complied with.

A velocity of 30 miles an hour to be considered the standard or minimum velocity – and this attained without requiring the piston to travel at a greater rate than 280 feet per minute.

The Engine to be of such dimensions and form as to maintain without difficulty with a pressure of steam in the boiler not exceeding 50 lbs the square inch a force of traction equal to 800 lbs upon a level at 30 mph.

The weight of the engine exclusive of the tender but in other respects supplied with water and fuel not to exceed 10½ tons and if above 8 tons to be carried on six wheels. The width clear between the rails will be 7 ft, the height of the chimney as usual. All materials and workmanship to be of the best description and except when modifications may be necessary to comply with the conditions stated or for the purpose of improvement to be similar to the same parts of the best engines now used on the Liverpool & Manchester Railway.

Drawings of the proposed Engines to be submitted to me as the Engineer of the Company before Execution and if during execution any material alteration is proposed it will be necessary that I should have the opportunity on the part of the Company of objecting to it if I should consider it an experiment not worth the making. I beg to request an early reply to this communication and to repeat that the object of the Company is by a liberal order limited by as few conditions as possible, the time of delivery being the most important to afford Manufacturers an opportunity of making those experiments which the specific nature of the Orders generally given my have hitherto prevented.

The 1st March 1837 is the period by which we hope you would be enabled to complete the order.

Brunel's requirements are strangely limiting considering that the standard gauge Liverpool & Manchester engines at this time were running at 30 mph with a piston speed of 504 feet per minute. Brunel's weight limitation was also extremely restrictive and less than most standard gauge engines on other lines, while to comply with the piston speed the wheels had to be made large. The largest wheels on contemporary standard gauge locomotives were 5 foot six inches in diameter, but in order to keep the piston speed within Brunel's limitations the first GWR engines had to have wheels of seven or eight feet in diameter. Large wheels were heavier, so, in order to keep the weight down, boilers were kept small and thus had little or no reserve of steam – another serious oversight made by Brunel.

Most of the engines produced to these specifications proved to be rather unconventional vehicles, especially the 2-2-2 *Ajax* which had all its wheels, including the 10-foot-diameter driving wheels, built up from solid plates instead of spokes. Other curiosities, admired by Brunel because of their novelty, were *Hurricane* and *Thunderer,* built by R. & W. Hawthorn under a patent of T. E. Harrison. They were articulated to overcome the weight problem, with cylinders and driving wheels of 0-4-0 configuration in front of the boiler and firebox which was carried on six wheels; being an entirely new concept, the idea naturally appealed to Brunel. It also had a water space dividing the firebox into two, the fire being fed through two doors. Unfortunately one fire tended to die down when the other was burning fiercely.

On 6 February 1838, Brunel wrote to Harrison:

I congratulate you upon the success which has attended the trial of your engine. I am puzzled as to the name – I am much obliged to you for the offer of naming it after me but it is quite inadmissible. I do take considerable credit to myself for having at once adopted what I believe will prove a most important change in construction and I shall be very much gratified if hereafter when the success becomes equally evident to all my name – my early 'patronage' as you call it – is remembered but I have no pretence to anything beyond this and my name might mislead.

I should prefer a distinct class of names which if possible should in some way have reference to their peculiar character and at the same time if possible which would admit of the addition of your name in speaking of them – I should propose that the first should bear your name simply but that I think as we have others making, & which I hope could follow, it would be a pity to fix your name until we determine which is the best of the three and in the meantime it will be called 'Harrison's engine'.

You ought to have sent me some details – does the gearing make much noise – how does the boiler work? I am very sorry that you have no reversing motion as it will prevent my venturing upon a greater speed as I otherwise should – at all events until I have ten miles clear run will not be until before the opening of the whole – I trust you have applied good brakes and then there will be *plenty of steam.*

I imagine there must be some screen to protect the engineman or he will be cut to pieces by the wind – have you arranged anything – and lastly let me call your attention to the appearance – we have here a splendid engine of Stephensons it would be a beautiful ornament in the most elegant drawing room and we have another of Quaker-like simplicity carried even to shabbyness [sic] but probably as good an engine but the difference in the care bestowed by the engineman, the favour in which it is held by others and even oneself not to mention the public is striking – a plain young lady, however amiable, is apt to be neglected – now your engine is capable of being made handsome and it ought to be so.

13 June 1838, he wrote to T. E. Harrison regarding the patent locomotive *Thunderer*:

With respect to the additional wheels to the engine with a tender tank. I think it best provided it is all securely attached to the engine frames but this as well as the alternative necessary for the pumps you really must take in hand at least through Messrs Hawthorn as neither I or anybody about me have time to attend to it which must be the case if made in London and let me beg of you to expedite the work if you have any wish that your plan should have fair trial on our line and that in spite of all my exertions anybody should not be prejudiced against them pray get it to work. A few weeks ago our Directors were willing to wait before we ordered more engines and looked forward to having many of yours now I am pressed every day to order at once for the next two years supply. This is very annoying to me but must be still more so to you and yet I have no answer to give but cannot run the engine to do any work even if we had Maudslay's gearing neither the tank nor the pumps would allow us. And these you must take in hand immediately. As to our using a tender I don't exactly see how as we have no tender to spare and are short of them as it is.

Pray let me hear that you are getting on with them.

Thunderer arrived 6 October 1838. Brunel wrote to Harrison on 25 October 1838:

You will be glad to hear that we are not likely to have any difficulty in raising the steam of the 'Hurricane' and I presume therefore with the 'Thunderer' whose boiler we used. I applied my nozzle to the blast pipe and we ran steadily at 45 miles an hour with a light train certainly as the steam was blowing off plentifully the whole way – there is still room for some improvement with blast pipes and I think it will do famously.

The engine wanted power however and I have not yet ascertained the cause – with that train (two 6-wheel and two 4-wheel and a truck) we ought to have gone more than 45 mph. The 'Thunderer' will be out in a few days.

Hurricane, like *Thunderer*, was also withdrawn in December 1839. Although it attained 60 mph on a test run, it was so light that it lacked the power to pull a useful load. Delivered on 6 March 1838, it was a complete failure and ceased work in December

1838, the last straw being when its tender fell off its chassis and caused considerable damage.

Ajax, the Mather, Dixon 2-2-2 engine which was notable in having 10-foot-diameter driving wheels built up with solid plates instead of spokes, was delivered 12 December 1838. Brunel seems to have been pleased with her, for on 23 April 1839 he wrote:

> Within the last 10 days I have had several opportunities examining the working of the Ajax – all the gearing of the engine appears to me to work very smoothly indicating sufficient strength and good workmanship. There is no priming & plentyfull [*sic*] supply of steam with a train consisting of three different railway carriages three loaded carriage trucks & a horse box – an average speed of 30 miles per hour was maintained exclusive of stoppages and it appears to me that when the engine was better known & understood by the engineer that the performance would be improved as it that experiment there was an excess of steam but the engine appeared throttled. During the high gales of last week the engine has been much more delayed than the others but the comparison has been made only with the two Stars, the Atlas & Lion engines of greater power – and which have been longer running – to what extent this delay is to be attributed to want of power or to increased resistance from the surfaces exposed I am not yet prepared to say but from the performance of the engine in fair weather I think that in power and speed it will be about on a par with the Vulcan – large engine while the workmanship is very superior and the consumption of coke singularly small.
>
> There is some trifling alterations required in unimportant parts of the engine but which can be better done when it is thrown off work for a day or two, at present it is necessarily in constant use.

Evidently she was not as good as Brunel first believed as *Ajax* ceased work by June 1840.

Sir Daniel Gooch in his *Memoirs* summed up Brunel's failings in the sentence, 'One feature of Brunel's character, and it was one what gave him a great deal of extra and unnecessary work, was that he fancied that no one could do anything but himself.' It was also short-sighted of Brunel not to have sought a tender from the Rennie Brothers to build locomotives, as their engineering works at Blackfriars was only a couple of miles from his office in Duke Street. Their standard gauge engines of 1838, as supplied to the London & Southampton Railway, would have almost fulfilled his requirements: they weighed 11 tons, reached 40 mph or more, and at 30 mph their piston speed was some 458 feet per minute.

Surprisingly for a brilliant engineer, Brunel seemed uninterested in the pressing matter of the standardisation of parts. On 6 September 1836, he wrote to J. Grantham of the locomotive builder Mather, Dixon, Liverpool:

> With respect to the buffers it is immaterial to me how they are placed between the engine and tender as I expect for the great variety in the construction that every engine must always be worked with its own particular tender which as it may be made at a later period may at the other end be adapted to the carriages as they may ultimately be determined on. At the front of the engine I shall instantly adopt as I expect to do with all the carriages high buffers 3ft 9in will be a very good height for this – can you adapt this to your present planning?

Another unwise habit of Brunel's practice was his writing to manufacturers asking them to build more locomotives before finding out whether the designs demonstrably worked.

Quite a contrast to Brunel's poor locomotives was Robert Stephenson's 'beautiful ornament' *North Star*, originally intended for the 5 foot 6 inch gauge New Orleans Railway and delivered by barge at Maidenhead on 28 November 1837, almost six months before the railway opened to that location. *North Star* proved most successful and ran until January 1871.

On 22 June 1836, Brunel wrote a letter concerning the Wharncliffe Viaduct to the contractors Messrs Grissell & Peto:

Gentlemen – just returned from Hanwell – observed that by far the largest proportion of the bricks upon the ground and actually in use were of a quality quite inadmissible ... I examined the bricks on Monday last and gave particular orders to your foreman Lawrence respecting which order I find he has neglected ... I must request that he be immediately dismissed.

Brunel did not always treat his contractors fairly. Hugh and David McIntosh particularly suffered and failed to receive their just deserts until an incident in 1865. Brunel interpreted words in their contract differently from others, 'coarsed rubble' to him meaning well-cut ashlar. When one embankment collapsed he insisted the space be filled with arches, declining to pay the additional cost. Unfortunately he was the arbitrator between the GWR and the contractor.

Furthermore, Brunel sometimes was not tactful with contractors, expecting them to make alterations at their expense or ordering their employees to do a task instead of asking the man in charge to see that it would be done. Brunel's prickly policy did him no good in these business relations, and the better contractors, who had full order books, would not work for him, especially as the contracts typically stipulated Brunel as the sole arbitrator. This attitude of the better contractors meant that tenders went to small or second-rate firms, some of which tried to do a second-class job, or even went bankrupt. William Ranger was one of these smaller contractors who carried out work on the GWR, including Acorn Bridge, a double-arch construction west of Shrivenham, and carrying the railway over the Wilts & Berks Canal, the River Cole and the Swindon to Oxford road (now the A420). On 3 June 1837, Brunel wrote a 'gee-up' letter to Ranger:

It was with great astonishment and regret that I observed when last at the Acorn Bridge that nothing was being done towards the long delayed erection of the steam engine. I cannot for one moment suppose that anybody would pretend to get the excavation for this bridge without some means of pumping and I see no preparations for any other means than that of an engine and there appears as regards this some cause of delay which is not communicated to me and unless I have immediately a clear and satisfactory explanation of all the present circumstances of your future plans and unless I can be satisfied that these plans are not only efficient but will be immediately carried into execution I shall wait no longer but without delay take all those steps which I consider necessary to regain a portion of the time which has been negligently wasted and proceed with the work in such a manner as I may find necessary. I regret being driven to this decision but the monumental dilatoriness of your proceedings and latterly by the apparent abandonment of all attempts to proceed leaves me no alternative. I shall feel obliged by an immediate reply.

Eventually Ranger was dismissed. He sued the GWR for non-payment but lost the case as his contract stated that Brunel had absolute power over his employment.

At the first meeting of the Bristol & Exeter Railway on 2 July 1836 its engineer, Brunel, anticipated it opening Bristol to Taunton and Exeter to Cullompton in the spring of 1838, and that the intervening gap would be filled by the end of 1839. This estimate proved ambitious as it did not reach Taunton until 1 July 1842, though the remainder to Exeter opened 1 May 1844.

In July 1837, Brunel received a letter from a twenty-year-old man, Daniel Gooch, stating that he was available for work, that he had experience working for Robert Stephenson and favoured the broad gauge. He first met Brunel on 9 August 1837, was promptly engaged in employment, and on 14 August 1837, ten days before his twenty-first birthday, he started work at the GWR engine depot at West Drayton. The first locomotives arrived that November.

Henry Booth, secretary of the Liverpool & Manchester Railway, sent a letter to the GWR directors complaining that the GWR had been poaching its engine drivers and proposed that all companies paid the same rates. Brunel issued his thoughts on the matter to the GWR directors on 19 September 1837:

My Dear Saunders,

I am sure that the Directors and you know me sufficiently to have been able to reply confidently to the letter which you have received that neither I nor anyone acting by my direction could have been guilty of tempting men from the employment of other companies.

Our Superintendent of Engines, Mr Gooch, who has been in the North to ascertain the state of progress of our engines and to ensure that proper enginemen should be ready to work them happened fortunately to return yesterday and I have had an opportunity of learning from him exactly what has been done.

Of the six men that he has engaged from the neighbourhood of Liverpool and Manchester and Birmingham one only is in the employment of either of the companies and this man (John Liver I believe) had given notice of leaving long before Mr Gooch was even in my service and he came to Mr Gooch recommended by an Officer of the Liverpool and Manchester Railway.

No temptation or inducement of any sort has been held out to men in the employment of this company or the Grand Junction but on the contrary it has been plainly stated to men who have offered themselves that such proceedings would not be sanctioned connected as the enginemen on their Lines are with the manufacturers who are making our engines it is highly probably that the circumstances of good men being wanted to accompany their engines has been a subject of conversation amongst the men and that they may have been asked even to make inquiries and it is a natural consequence that the men should avail themselves of the supposed demand for their services either as an excuse if they wanted to leave or in hopes of adding to their own self-importance if they remained. This is the natural consequence of our wanting men of a class who must of course come from the neighbourhood of these railways and in some way or other and at some time have been connected with them but I repeat that neither I nor any Agent of mine to my knowledge has directly or indirectly induced any man to leave his employment in their service to enter ours.

With respect to the amount of wages to be given I have been guided by the London & Birmingham and believe it will be found necessary to give higher wages here than in the north but upon this point I will be prepared on Thursday. As Mr Booth refers to the advisability of agreeing to give the same wages on all railways I think his proposition must amount to this. I will just observe that I think if would be very improper to set such an example to the men as combination we could hardly complain if in their turn they were to 'agree' also as to a minimum and one company that might think it necessary to secure more careful men would be precluded from doing so.

Brunel encountered problems with his first iron bridge at Uxbridge Road, Hanwell. The iron frame was carried on brick abutments and two rows of columns. One beam broke during construction in 1837 and was replaced, only to be complicated further when another broke in 1839.

Although in the early days special designs were made for every one of the bridges over and under the railway, with the rapid expansion of the broad gauge system the number of bridges designed became so large that Brunel had a set of standard drawings prepared and engraved ready to be used in various suitable situations. Contract drawings were made by adapting the particular circumstances of each case to the standard template which was most applicable.

Brunel liked ballast on his bridges and because there was no change in the nature of the support given to the rails, no concussion was caused on a train entering or leaving a bridge. The ballast absorbed the vibration of a train, and in the event of a derailment helped to prevent rolling stock ploughing through the decking. The extra material required to support the weight of the ballast was more than compensated for by these advantages. Timber flooring on iron bridges was generally laid diagonally on wrought-iron girders, placed not at right angles to the line but obliquely in order that two wheels of the same axle of an engine or heavy wagon might be on different cross girders. This arrangement allowed the cross girders to be of less strength and thus allowed a saving in their cost and weight.

Charles Saunders, the GWR's secretary, was an intimate friend and the closest to Brunel. On 3 December 1837, under very great strain as the opening of the GWR approached, Brunel wrote:

My Dear Saunders,
A hint or two from the other end is useful now and then to remind me of what, however, I am fully sensible of and always thinking of, your exceeding kindness in relieving me of everything you possibly can, and still more strongly shown in your silence, and the absence of complaints.

In my endeavour to introduce a few – really but a few – improvements in the principal part of the work, I have involved myself in a mass of novelties.

I can compare it to nothing but the sudden adoption of a language, familiar enough to the speaker, and, in itself, simple enough but, unfortunately, understood by nobody about him; every word has to be translated. And so it is with my work – one alteration has involved another, and no one part can be copied from what others have done.

I have thus cut myself off from the help usually received from assistants. No one can fill up the details. I am obliged to do all myself, and the quality of writing, in instructions alone, takes four or five hours a day, and an invention is something like a spring of water – limited. I fear I sometimes pump myself dry and remain for an hour or two utterly stupid.

As regards the Company, I never regretted, one instant, the course I have taken … And, as regards myself, if I get through it with my head clear at all, I shall not regret it, but I certainly never should but for your kindness, and the corresponding forbearance and kindness of our Directors.

I have spun this long yarn, partly as a recreation after working all the night, principally to have the pleasure of telling a real friend that I am sensible of his kindness, although he hardly allows me to see it, and partly because I wish you to know that if I appear to take things coldly it is because I am obliged to harden myself a little to be able to bear the thought of it.

If ever I go mad, I shall have the ghost of the opening of the railway walking before me, or rather standing in front of me, holding out its hand, and when it steps forward, a little swarm of devils in the shape of leaky pickle-tanks, uncut timber, half-finished station houses, sinking embankments, broken screws, absent guard plates, unfinished drawing and sketches, will, quietly and quite as a matter of course and as if I ought to have expected it, lift up my ghost and put him a little further off than before.

What a note I have written! Well, I do not think it is altogether wasted. I hope you will think the same.

Brunel had originally planned for a tunnel at Sonning, but then decided that a cutting two miles long and with a maximum depth of 60 feet would be preferable. The contract was let to William Ranger, but he worked so slowly that in August 1838 he was replaced.

7

A STEAMSHIP FOR THE
ATLANTIC, *GREAT WESTERN*

Thomas R. Guppy, one of the Bristol directors who was a career engineer turned to the business of sugar refining, took Brunel's flippant steamship suggestion seriously. The two, together with a semi-retired Royal Naval officer, Christopher Claxton, formed a committee. Claxton and Bristol shipbuilder William Patterson sailed on various British steamships and recommended constructing a much larger steamship than any built to date.

The received wisdom was that as small ships could not carry enough coal to cross the Atlantic, engines were only suitable for use on windless days. It was wrongly assumed that if a small ship could not carry enough coal, neither could a large vessel. Actually a vessel's resistance as it was driven through the water did not increase in direct proportion to its tonnage. Resistance increased as a square of the dimensions of the width and depth of the hull, whereas tonnage increased as the cube of these dimensions. This meant that the need for fuel to overcome resistance increases at a lower rate than the capacity of a ship, so that the fuel bunkerage may be decreased; a very large ship then would have space for coal and machinery and still have plenty of space for freight or passengers. Claxton and Patterson calculated that a ship of

1,200 tons and 300 horsepower loaded with 580 tons of coal could average between six and nine knots and cross the Atlantic in less than twenty days westwards and thirteen days eastwards, thus halving a sailing boat's time.

Brunel also realised that with large ships longitudinal strength becomes an important variable in order to prevent bending when the centre is unsupported. His solution was to use box girders, as on his bridges.

On 3 March 1836, the Great Western Steam Ship Company held its first public meeting. It was planned initially to construct two ships according to Claxton and Patterson's size and design, but later decided upon to build just one even larger vessel of 1,320 tons and with a 400 horsepower engine. As managing director, Claxton was responsible for day-to-day operations; Patterson assumed the lead for the hull and Brunel for the engine. Whenever railway business brought Brunel to Bristol, the committee of Brunel, Claxton, Guppy and Patterson convened to discuss business – this happened approximately once a week throughout the project. In the event, Brunel did not contribute much to the building of the *Great Western*, though with his knowledge of bridge stresses he recommended adding extra bolts and trusses to increase the ship's longitudinal strength. Brunel desired larger cabin windows in the stern but was reminded by Claxton 'that there was water outside which was sometimes very uneven in its surface, and unlike the generality of lawns'. The keel of the *Great Western* was laid without ceremony in June 1836.

In August 1836, the controversial Dr Dionysius Lardner addressed a meeting in Bristol of the British Association on the subject of steamships crossing the Atlantic and tried to put a spoke in Brunel's wheel. He produced calculations which showed that the idea was physically quite impossible. 'Take a vessel of 1,600 tons provided with 400 horsepower engines. You must take 2 ⅓ tons for each horsepower, so the vessel must have 1,348 tons of coal. To that add 400 tons, and the vessel must carry a burden of 1,748. I think it would be a waste of time, under all the circumstances, to say much more to convince you of the inexpediency of attempting a direct voyage to New York.'

Pressing ahead in defiance of its critics, on 19 July 1837 the *Great Western* was floated out of the building dock at Bristol and set sail on 18 August 1837 for London to have its engine fitted. This was built by Maudslay, Sons & Field, chosen as the Brunel family had been connected with the firm now for almost four decades; Thomas Guppy, Brunel's colleague, had also worked there for five years. The firm provided a 400 horsepower, two-cylinder side-lever engine capable of driving either just one paddle wheel or both. Joshua Field, an eminent engineer at Maudslay's, designed a system of engine cams that improved steam economy and initiated more efficient double-storey boilers. Field also developed the spray condenser which converted some of the engine's used steam back into fresh water – this limiting scale deposits enough to permit the boilers to be fired continuously across the Atlantic without having to be emptied for cleaning. Brunel realised that a small amount of steam in a cylinder would continue to heat and expand under the piston's compression, so he invented a variable cut-off to reduce the amount of steam first entering a cylinder, thus making more economic use of the engine.

Following the engine being installed in London, and the rest of the wooden-hulled vessel being fitted out there, on 31 March 1838 the *Great Western* left Blackwall for Bristol to begin her maiden voyage to New York. Both Brunel and Claxton were on board for the voyage. All seemed to be in high spirits until the ship was off Leigh-on-Sea at the mouth of the Thames, where a serious fire broke out in the engine room and the felt insulation around the boilers ignited.

The deck caught alight and flames reached the top of the funnel, thus making it difficult to reach the source of the fire. Claxton grabbed a leather hose and hauled it to the forehatch and from there projected water on to the fire. Brunel, attempting to assist, lost his footing on a burned ladder rung, causing him to fall about 20 feet heavily onto Claxton. There Brunel lay unconscious, face down in a puddle. Claxton, who had inadvertently saved his life by providing a soft landing, then dragged him from the puddle and yelled for a rope. Brunel was hauled up to the deck with a dislocated shoulder and broken leg. Wrapped in a sail, he was quickly lowered into one of the ship's boats and rowed ashore to a farm on nearby Canvey Island, where he had to remain for several weeks.

The captain, Lt James Hosken, contemplated lowering the lifeboats and taking off his passengers, but decided instead to beach the *Great Western* on Chapman Sand; there the crew managed to break through the decking into the engine room and douse the flames. On the next tide she was refloated and continued to Bristol, the only damage consisting of burned felt and charred wood. On 3 April 1838, Brunel dictated a long letter to Claxton reporting the ship's satisfactory performance, stating: 'I hope the Vessel will be a long way on her voyage to New York before I could be in a state to go onboard again.'

The PS *Great Western* departed for New York on schedule on 8 April, offering only a first-class service, no accommodation being made for second-class or steerage passengers. Unfortunately, the fire scared away potential passengers and it only carried seven travellers at thirty-five guineas apiece, instead of the expected fifty-seven. The crew also numbered fifty-seven, including twenty-four seamen and fifteen stokers; in addition to these stokers feeding the furnaces, trimmers were required to transport the coal by basket and wheelbarrow from the holds at the bow and stern. During the voyage the engines were only stopped thrice for minor adjustments and to take soundings on the Newfoundland Banks.

Life on board a steamship proved different from a sailing packet. All day and all night the engine pounded away; smoke and smuts needed to be avoided; and the engine lubricant, derived from animal fat, had an odour, unpleasant to anyone suffering from seasickness. Heat from the boilers caused the tar caulking the deck to bubble. New York was eventually reached on St George's Day, 23 April, the fastest crossing yet from England to America – done in only fifteen and a half days. Of the original 660 tons of coal in her bunkers, 203 tons remained, leaving a satisfactory margin of safety – Dr Lardner's calculations had proved incorrect. The rival ship PS *Sirius* had taken nineteen days and only had fifteen tons of coal remaining in her bunkers.

An American reporter recorded the arrival in New York:

> The approach of the *Great Western* to the harbour, and in front of the Battery, was most magnificent. It was about four o'clock yesterday afternoon. The sky was clear – the crowds immense ... Below, on the broad blue water, appeared this huge thing of life, with four masts and emitting volumes of smoke. She looked black and blackguard ... rakish, cool, reckless, fierce, and forbidding in sombre colours to an extreme. As she neared the *Sirius,* she slackened her movements, and took a sweep round, forming a sort of half circle. At this moment, the whole Battery sent forth a tumultuous shout of delight, at the revelation of her magnificent proportions. After making another turn towards Staten Island, she made another sweep, and shot towards East river with extraordinary speed. The vast multitude rent the air with their shouts again, waving handkerchiefs, hats, hurrahing.

Sixty-six passengers made the return voyage, done in fifteen days, but the round trip had lost the company nearly £4,000. She made four more return trips that year and earned a small profit. She carried out a further sixty-seven Atlantic crossings in eight years and was the first ship to hold the Blue Riband, her best crossings being thirteen days going westwards and twelve days and six hours eastwards.

The accident on board the *Great Western* had certainly shaken Brunel, for on 11 May 1838 he wrote to his friend Thomas Guppy:

> Thank you very much for your inquiries – I was much grieved to receive your letter a confirmation of the report which had reached me through a friend in Paris.

I am not particularly well in body or mind – I don't get away – I am still lame in the left foot and my back is weak. I don't write this letter without leaning back to rest and in consequence of the state of my stomach I am nervous anxious & unhappy – in fact blue devilish.

An infinite number of thoughts crowding in upon me requiring attention & thought – all in arrears & I am quite incapable of getting through them – everything seeming to go wrong – we talk of the 30th [May] for opening [the GWR] and now everybody believes it but me – I suppose I want a dose of salts.

On 19 May 1838 he wrote again to Thomas Guppy:

Thank you very much for your letter – I wish it had been yourself instead I should like a good skeming [*sic*] chat such as we have had sometimes.

What is the truth in these conflicting accounts of the GW steamship – by this time probably you know the best information. I am most anxious – 15 days at all events seems the maximum and under the circumstances it is as you say an admirable voyage – beyond all expectation good – but again why are they not back? If she has arrived today I suppose Claxton will have sent parcel there being no post. [The PS *Great Western* dropped anchor at the mouth of the Avon on 22 May 1838, having taken fourteen days from New York].

The GWR goes well – but what with the circumstances of my not having seen much of it and what with my nervousness I am very frightened I can hardly say of what I am particularly anxious. Hammond and others have worked most zealously and have all my instructions but still one's own eye is the only one after all – and I dread the trial of each part of the line as it comes into operation – the engines are running today from Maidenhead to Hanwell and we had trips from Paddington up to the crossing of the Thames Junction. On Tuesday we intend running with some Directors from Hanwell to Maidenhead in hope to be open the 3rd [June] – we still talk of 30th [May].

I thank you very much for your advice. I wish I could follow it – and still more do I thank you for your kind offer of assistance and so sincere do I believe it that I should accept it but that on consideration of time is too short the press too immediate for any remedy now – it would require more thought of method to avail myself of the assistance of a person new to those about me than to struggle on as it is – I need hardly say that the subject of a former conversation has frequently forced itself upon me but I have not allowed myself to think over it until I find myself better fitted for cool consideration.

Certainly the kindness and consideration of all around me, our directors especially, is almost enough to relieve my mind from anxiety and that is saying much indeed but it's uphill work still to keep my spirits up – you will find this a desperate long scrawl and rather over much about self but it's a recreation – next to a pleasant chat – after many a dozen dull letters I have had to write with still a fearful heap of arrears by my side.

In due course the PS *Great Western* was requisitioned by the government in 1855 for use as a troopship during the Crimean War. Following the cessation of hostilities a refit was deemed uneconomic and she was broken up at Vauxhall in 1857, Brunel visiting her there.

Before the opening of the GWR, the Bristol & Exeter Railway was proposed to extend the GWR westwards. A prospectus was issued on 1 October 1835, with Brunel appointing William Gravatt to carry out a survey in November 1835; the Act was passed 19 May 1836.

William Gravatt was a long-standing friend, having been appointed back in November 1826, when only aged twenty, to assist and alleviate stress on an overworked Brunel in the Thames Tunnel project. Gravatt died aged sixty, poisoned by an accidental overdose of morphia administered by a nurse. The Bristol & Exeter Act was followed by the Cheltenham & Great Western Union Railway Act on 21 June 1836, this line leaving the GWR at Swindon. Brunel was not keen on his position of engineer of the latter, writing: 'None of the parties are my friends, I hold it only because they can't do without me – it's an awkward line and the estimate's too low.'

Brunel's increasing fame as an engineer meant that he often received letters from parents requesting a pupilage for their son or protégé. On 16 January 1836, he wrote to the young W. G. Owen (who became Chief Engineer of the GWR in 1868):

In consequence of Mr Bennett's [Brunel's clerk at Duke Street] strong recommendation I authorised him to write to you on the subject of your being employed upon the Great Western Railway ... but as I have not the means of judging your capability I must explain to you the terms upon which you, or any other gentleman, must enter the service of the Great Western Railway. The sub-assistants must be considered as working entirely for promotion, their salaries and their continued employment depend entirely upon the degree of industry and ability I find they possess. Their salaries commence at £150 p.a. and may be increased progressively up to £250 and perhaps in some cases £350 p.a. They must reside on such part of the line as required, consider their whole time, to any extent required, at the service of the Company and will be liable to instant dismissal should they appear to me to be inefficient from any cause whatsoever and, more particularly, to consider themselves as on trial only. If these conditions appear to you encouraging you will come to Town immediately and call at my office, 18 Duke Street, Westminster.

Despite these strong words, in practice a man had to be inefficient over a period before Brunel sacked him.

Most of the engineers trained by George Stephenson were employed by other lines and so Brunel was forced to train his own. In April 1836 he wrote:

Hudson has left me, Stokes is an imbecile and I am rather in a mess, but S. Clark is a god-send and Hammond is a good fellow and very useful and has come up and endeavoured to put my accounts in order.

Although for much of his life Brunel was financially well-off, he achieved it mostly from investments rather than from his professional work. One of his guiding principles was to invest in his own schemes and thus share in the financial risks. In April 1836 he wrote in his journal:

One thing however is not right; all this mighty press brings me but little profit – I am not making money. I have made more by my Great Western shares than by all my professional work – *voyons* – what is my stock in trade and what has it cost and what is it worth?

Early in 1836 a sub-assistant, Harrison, working at the Hanwell Viaduct under J. W. Hammond, was sent a letter by Brunel:

My Dear Sir,

I am sorry to be under the necessity of informing you that I do not consider you to discharge efficiently the duties of assistant engineer and consequently, as I informed you yesterday, your appointment is rescinded from this day. A great want of industry is that of which I principally complain and thus it is entirely within your power to redeem the situation. I shall have no objection to you continuing on trial at the Bristol end of the line and I sincerely trust that you will see the necessity of making greater exertions ... and that at the expiration of a month I may be able – with justice to the Company – to reinstate you in as fine an opportunity of advancing yourself in your profession as a young many could possibly have.

This letter was duly sealed, but before its dispatch Brunel received a bill for a circumferentor, an instrument which preceded the theodolite. Brunel had told Harrison that no surveyor should be without one and Harrison, believing it was for the Company, sent the bill to Brunel. The latter, furious at this perceived insolence, broke the letter's seal and added:

> PS You have acted with reference to this in a manner which I do not choose to pass over. It indicates a temper of mind which excludes all hope of your profiting from the new trial I had proposed. You will please consider yourself dismissed from the Company's service on receipt of this letter.

The river crossing at Maidenhead posed a problem to Brunel. It had to avoid obstructing the towpath and be of sufficient height to allow sailing barges to pass below. This limited Brunel to just one river pier and to avoid increasing the 1 in 1320 gradient the two arches would be required to be flatter than any designed so far. Each had a span of 128 feet with a rise of only 24 feet 3 inches to the crown – his detractors were convinced it would collapse. When on 1 May 1838 the contractor Chadwick eased the centrings, three courses of the eastern arch dropped half an inch, but the western arch stood firm. It was not a design fault but caused by the contractor, Chadwick, who admitted that he had eased the centrings too early before the Roman cement had properly set. Although the centrings were eased on 8 October 1838 they were not removed, and Brunel enjoyed his critics supposing that the centring was supporting the bridge. Chadwick repaired the damage and the architecturally pleasing bridge still stands to this day.

The Thames Commissioners received a complaint that the construction of this bridge blocked both waterway and towpath, and wrote to Brunel without first checking the facts. Hurt by this presumption, on 29 August 1836 Brunel wrote a fiery reply:

> I have taken all due steps to ensure the security of the Navigation during the construction of the bridge – and at considerable expense and in a much more complete manner than I believe you yourselves would have provided had you

been executing the work. I have designed the largest arch of any bridge above Southwark for the express purpose of including the whole of the present Navigation channel and the wide towing path under one arch. I have done what I suspect I will be severely criticised for, namely a 2-arch bridge, for the purpose of throwing the only pier required on an existing island. I believe this will be the only instance on the entire river of a bridge being built without in the slightest degree altering still less injuring the navigation. After doing all this – far more than we could have been compelled to do – unasked – it is rather vexatious to be interfered with in the manner we have been.

I hope you will put all right and restore the good feeling you feel disposed to cultivate. I have explained what I intend and cannot be responsible for the statements others may choose to make. The bridge has been designed peculiarly with regard to the navigation and without consideration of expense on the part of the railway. If the commissioners are discounted and troublesome there is still time for me to save £10,000 and still build a bridge as they can find no fault with – but it would be by no means as convenient as the one at present intended.

Despite this display of a lack of good faith, Brunel's demand and reputation continued to grow. On 3 October 1836, he wrote to a parent of one aspiring pupil, Henry Sale of Merthyr:

I do take pupils – or rather I have been driven to take them – I have no room now and I shall have none for twelve months – my terms are 600 guineas ... P.S. I would not take a pupil without six months notice.

By 1843 he charged a premium of £1,000.

On 17 April 1846 he wrote to J. A. Whitcombe:

I take pupils but rather as the exception than the rule: that is, I do not seek it and rather raise difficulties than encouragement, and amongst other difficulties which I create is the premium I charge. I don't know whether it is high compared with what others charge but I know I should think it a great shame if I were a father wanting to put my son with an engineer. I charge £1,000 that is to say £550 on entering and £150 for three years, and one year for nothing making in all four years during which time I profess to take no trouble whatsoever about the youth. He has all the opportunities which my office of course give him, and if he turns out well gets employed in responsible situations which improve him. Do not be induced to expect more for your son but at the same time I cannot deprive myself of the opportunity of saying that it would at all times and therefore in so important a matter as your son's welfare afford me *great pleasure* to forward any views you have.

Brunel envisaged the Cheltenham & Great Western Union Railway as a link to South Wales, where in 1834 he had become friends with Anthony Hill, the Taff Vale ironmaster who he had met in connection with the ironwork required for the Clifton Bridge. Hill sounded out Brunel regarding a line from Merthyr to Cardiff which could convey coal and iron from the Taff Vale to Cardiff for shipment. He met with Sir Josiah Guest, who became the first chairman of the Taff Vale Railway at 6.00 a.m. on 12 October 1834, Lady Charlotte Guest recording in her diary that it was 'a *very* early meeting'.

8

THE GREAT WESTERN
RAILWAY OPENS

Before the opening of the section to Maidenhead, Brunel tried Wheatstone & Cooke's new telegraph. It used five wires, each insulated with cotton and gutta-percha carried in an iron pipe above ground beside the rails. Powered by voltaic batteries, a display of five magnetic needles as dials could be moved through various permutations to signify all the letters of the alphabet and the numbers one to ten. Unfortunately a deterioration of the insulation marred its working operation, but an improved version of the telegraph requiring only two wires suspended from cast-iron standards proved better.

The line was opened from Paddington to Maidenhead on 31 May 1838 when 300 guests, were carried in two trains. The first headed by *North Star* ran at an average speed of 28 mph but returned at 33½ mph, Thomas Guppy foolishly walking along the roofs of the coaches as they travelled at 50 mph. When the line opened to the public on 4 June 1838 it was soon discovered that the Brunel-designed locomotives were quite hopeless and incapable of attaining a reasonable speed.

The coaches gave a rough ride, Brunel attributing this partly to poor ballasting below the longitudinal timbers' sinking and partly to the coaches' poor springing. The introduction of six-wheelers and a repacking of the road with coarser ballast caused some improvement.

Brunel's first coaches were four-wheelers, and unfortunately opted for a wheelbase approximately the same as the rail gauge – this fact led to very rough riding. Fortunately, even before the line was opened to the public, he decided to standardise by using six-wheel passenger coaches. These offered a much more comfortable ride; most of the four-wheelers were withdrawn within two months of the line opening.

Brunel's ideas of carriage design were as odd as those he had for locomotives. His first-class coach was 7 feet 6 inches wide down to the seating, but then narrowed to 6 feet in order to accommodate the 4-foot-diameter wheels which came outside the body. The 18-foot 6 inch-long bodies were carried on axles 10 feet apart – Far too short a wheelbase and at speeds of over 30 mph would have suffered hunting oscillation from side to side. Their seating capacity was only eighteen – less than a first-class coach on the London & Birmingham line.

Even before these odd coaches were actually built, and almost a year before the first length of the GWR opened, Brunel designed a far better carriage on six wheels and with the bodies directly above the 4-foot-diameter – yet he said that the main reason for his broad gauge was to have the body *within* the wheels! Although the GWR's 4 foot diameter wheels were heavier than other railways' 3-foot wheels, and the space allowed for a GWR passenger was slightly more and the height of the body greater, yet the gross weight per passenger was less overall.

	Tons	cwt	qr
A Birmingham first-class coach weighs	3	17	2
With 18 passengers at 15 to the ton	1	4	0
	5	1	2
Or 631 lbs per passenger			
A GWR first-class weighs	4	14	0
And with 24 passengers	1	12	0
	6	6	0
Or 588 lbs per passengers			

GWR 6-wheeled first-class	6	11	0
With 32 passengers	2	2	2
	8	13	2
Or 600 lbs per passenger			

The GWR directors received a tribute from an elderly clergyman from Banbury, who wrote:

> I do not want the little hair on my head rubbed off by riding in the London and Birmingham Company's low carriages. I would vastly prefer travelling the whole distance in one of those splendid carriages of the G.W. Co. than being transferred to one of those little pig-boxes of the London and Birmingham line.

Whenever Brunel travelled by train he evidently considered himself still at work; even at a speed of 50 mph he would work his way along the footboards from one coach to the next ascertaining its riding ability. The early carriages were made by stagecoach builders rather than engineers, and axles and bearings were frequently incorrectly designed. Brunel thus brought the Carriage & Wagon Department into being, when on 5 August 1842 he suggested to Charles Saunders, the GWR secretary:

> I have as you know very frequently called attention to the state of the different carriages which I find defective as I make a rule of always trying each carriage in the train I travel by. The result being that these carriages are taken off work and the defects rectified – generally also I have pointed out the cause of the defect and suggested the best mode of remedying them.
>
> Both the Clarkes [Seymour and Frederick, traffic superintendents at Paddington and Bristol respectively] have always shown the greatest desire to ask my advice as to the work to be done – and I believe are always anxious to adopt that advice – thus I have frequently seen what I have considered very injurious and un-mechanical things done – and I don't

think a mere coachbuilder however good is for the engineering of railway carriages – for instance the ironwork and fittings of our framework are by no means what they ought to be and one great defect which has annoyed us so long namely that awful thumping of the wheels when going fast I think would not have occurred had an ordinary millwright or engineer examined and received the wheels – after a great deal of trouble and very close examination I discovered the causes – an inequality in the thickness of the tire which threw the wheel out of balance – when the wheel is clean and new this would have struck the eye of an tolerably good mechanical workman when deliberately inspecting the wheel – and a large expense saved to the company. Besides 'thumping' I believe the defect is principally in a certain size of wheel and is rather peculiar to our railway as the 'thumping' taking our wheels of 4ft to the others a 3ft high and our high speeds about the same proportion will be about as 2½ to 1.

I think all the constructional parts of the carriages should be in some way under the engineering inspection and further – done at one place. I found a very bad carriage today – bad enough to stop it running to Taunton but I preferred letting it run to telling Clarke because I have found these carriages get repaired first at one [place] and then at the other and have seen extra-ordinary remedies when the physicians call – I think incompatible remedies apply and more than one – they get terribly patched up and a great deal of work is required and not very profitably.

I think all the wheels, axles, springs, oil boxes, buffers, heavy iron-work and bolts should be made (or examined, if delivered under contract) at Swindon by the same class of people as the engine work is – for it ought to be carefully made – to be taken into store there.

I believe that Swindon would also be the proper place for effecting any material repairs – small repairs ought to be effected at Paddington or Bristol if they always draw the parts ready made from the General Store because I would undertake them if the whole was under one management and a proper class of workman, the constant little doctoring now required would cease.

The thumping which puzzled us all and what has caused so many carriages to have something done to them really arose as I before explained from a defect which wd have caught the eye of an experienced workman – these dreadful motions which have shaken our carriages to pieces would never have arisen.

I discovered quite indirectly that the bearings of the axles having worn longer, the pattern of the brasses perhaps shorter, they were in the habit of using brasses ½ inch or more too short – no mechanic would have done that. There are fifty other points I could refer to but they all form the same thing – that a railway carriage requires engineering superintendence and that is quite impossible that Seymour Clarke can conduct such a department while Fredk I don't think is at all competent & altho' the former is very intelligent he has not had the mechanical practice that is required.

Let me know your views when we meet.

[PS] Of course you will not hint at my opinion of the Clarkes to anybody but think over what I have said.

To help him in his work Brunel kept a commonplace book noting any fact he discovered which might prove useful. It includes tables of rainfall and the differences of local time from London time. He noted species of grasses most suited for growing in various soils and the fact that the original cast-iron rails of the Hetton Colliery Railway had a life of ten to twelve years. The cast-iron rails of the Stockton & Darlington Railway were more durable by fifty per cent than the malleable rails.

In rolling iron rails, scale formed due to cooling during the manufacturing process, and this in turn caused lamination and splitting. The heavier the rail section, the greater the risk, so Brunel accordingly designed an inverted 'U'-shaped rail, light in section but also strong. His initial rail, at 43 pounds per yard, proved to be too light and this was increased to 62 pounds per yard west of Maidenhead. Throughout 1839, the Paddington to Maidenhead line was relaid with 75 pounds per yard rail.

On 29 May 1840, Brunel shared publicly his intentions to use a lighter rail for the new line. 'Each pound weight per yard represents £40 per mile. Our present rail is 18 to 19 lb more than the original and I suggest 10 or 12 might do and offset the cost of the proposed improved joint. I intend to lay half a mile beyond Boyne Hill Bridge with old rails as an experiment. If 55 lb is suitable it will save £24,000 on the whole line.' As a result of his experiment, a new standard of 62 pounds per yard was decided upon.

Brunel's rail had a higher strength to weight ratio than any contemporary rail. Sir John Wolfe Barry in *Railway Appliances*, published in 1876, said that Brunel's 62 pound per yard rail was equivalent to 75–85 pounds of normal rail on transverse sleepers. In addition to the saving on the original cost of the iron, incurred maintenance costs were lower, A.W. Gooch of the Oxford Division engineer's office calculated that maintenance costs for a baulk road was about £20 per mile cheaper annually than a cross-sleepered line, though when a line became mixed gauge annual maintenance costs soared by about £87.

Brunel did not exclusively favour the broad gauge. The Taff Vale Railway was incorporated by an Act of 21 June 1836 with Brunel as engineer; two of its first directors were also on the original GWR Board. Brunel recommended that its gauge be 4 foot 8½ inches. He later informed the Gauge Commissioners that he chose this gauge as being more suited to the many sharp curves he was obliged to adopt for the line's course. To cope with severe inclines of 1 in 19 and 1 in 22 he installed stationary engines – these continuing to work the section until 1864.

Brunel had made the mistake of making the GWR track rigid. George Stephenson used a permanent way, consisting of short rails secured in iron chairs fixed to stone blocks. These blocks sank under the weight of passing trains and imparted an up-and-down motion to rolling stock. Brunel believed riding could be improved by using relatively light rails secured to longitudinal timbers. His inverted 'U'-shaped rails were clever, as their wide flanges spread the load on the supporting timbers, while the crafty short portion of an edge of 1 in 10 centred vehicles on the track and prevented them yawing from side to side. To hold these beams to gauge, every 15 feet the Up and Down roads were linked by transom. These transoms in turn were fixed to the line's foundation by vertical piles 12 feet in length. These piles held the track down securely so that stone packing could be placed below the longitudinal sleepers supporting the rails. Brunel's track on longitudinal baulks used less timber than a cross-sleepered road.

Although longitudinal sleepers were not novel, the use of piles and a gauge of 7 foot ¼ inches certainly were. Brunel was being somewhat devious when he reported to his directors:

> I have no new plans of my own ... on the contrary, the result of the best considerations ... induces me at present to give preference to a slight modification of the oldest system of railway building used in England ... namely rails of simple form laid on longitudinal beams of wood ... I beg to repeat ... that in this system there is nothing new ... each part is old and has at some time or other been subject to experiment.

However, some of the packing below the longitudinal beams proved defective (Brunel was absent at the time on trials of the PS *Great Western* and was unable to spot this deficiency). This meant that the piles merely supported the timbers instead of holding them down. This led to a switchback road with rails supported by piles and consequent dipping in between.

Brunel attributed the poor riding partly to poor ballasting, with fine gravel below the longitudinal timbers, and partly to the coaches' poor springing. As an experiment, half a mile of track was re-laid with 18 inches of well-rammed coarse gravel retaining the piles, with an adjoining similar length similarly ballasted but without piles. As a result of this trial Brunel severed the piles below the transoms and used heavier baulks.

Before its laying, all timber was kyanised, a preserving process invented by Dr John Howard Kyan and patented by him on 31 March 1832; an Act of 26 May 1836 assigned his rights to the Anti Dry Rot Company. Brunel paid great attention to wood preservation. As early as 1835 he had been in communication with Michael Faraday as to the best method for testing the extent to which the kyanising solution penetrated into the wood. Brunel on several occasions kept the operation of preserving timber in the remit of the railway in order that it might be done thoroughly under his own supervision. Beside the Grand Junction Canal at West Drayton and at Thames Wharf, Maidenhead, Brunel set up pickling tanks where the timber was seeped for eight days in a solution of mercuric chloride. By 1840 kyanisation was superseded by immersion in creosote. Unfortunately, kyanised timber was inflammable. On 31 May 1848, a workman heated a large iron bolt to secure the timber bridge over the River Usk, and when driven home it set the whole structure alight.

George Gibbs, a GWR director, and Charles Saunders, the secretary, rode on the London & Birmingham Railway and found that the track laid by Robert Stephenson on stone blocks was little, if at all, better than that of the GWR – the bumps and jolts at the joints being very frequent and in some places very uncomfortable.

In due course the piles were severed and a heavier section of rail laid in place. Permanent way engineers thus discovered that a certain amount of resilience in the road was essential. However, Brunel's longitudinal sleepers proved superior to the cross variety, as in the event of a derailment, which was far from infrequent in the early days of rail travel, they would keep the rolling stock more or less in line.

Brunel was concerned about the ongoing issue of riding qualities and wrote to his friend Thomas Guppy on 31 July 1838:

When I received yours this morning I immediately determined to join the party but am nailed for Thursday and Friday – I fully hope however to be at Liverpool on Saturday morning and shall try hard to persuade you to cross again.

Now the points are – do the carriages run smoother and with less noise – it is a continuous bearing and exactly what is its construction – and what is the packing.

Does it not jump or creep under the wheel? If it's not held down I'm sure it must.

The Hanwell embankment has been particularly quiet – it was a false alarm – no foundation whatever.

We had a smash on Sunday night [29 July] – a return engine and train (no passengers) ran into some earth wagons – the engine (North Star) jumped off 6ft from the line *hard against* the other line into which we lifted her – in spite of long axles the crank axle is not strained *in the slightest* and the others only every slightly – but for the feed pipes being broken she could have worked home.

The fact that *North Star* was so little damaged in this mishap speaks very highly for the broad gauge's advantages.

Brunel improved rail safety by designing good signalling systems. Initially signalling was performed by hand signals given by railway policemen, but by 1841 Brunel had devised a better way. A disc and crossbar were set at right angles on a pole. When the crossbar was displayed, meaning that the disc was edge-on and virtually invisible, this indicated 'danger'. 'Clear' was shown by the disc being displayed when the pole was turned so that the cross bar was edge-on. Signals referring to Down trains had down pieces hanging from the end of the crossbar to differentiate from the plain crossbar for Up trains. On other railways, no signal indicated a general 'All clear', but that devised by Brunel gave a positive signal for 'All clear' and 'Danger'.

The Grand Junction Canal bridge at Paddington was designed by Brunel in 1838 and was the first he designed in iron. Built by Sherwood, the contractor for the brick-arched section of Bishop's Road Bridge immediately to the south, it was completed in 1839. The original railings of the canal bridge were removed in 1906 and replaced with high, brick parapets. The bridge was dismantled in 2004 and stored, pending re-erection about 200 yards up the canal from its original location.

During the summer of 1838 work began on Sonning Cutting east of Reading. Almost 2 miles in length and with a maximum depth of 60 feet, it was one of the largest excavations of its kind. It was dug by 1,200 navvies aided by some 200 horses. The heavy rain in November 1839 flooded the site and rendered the unfinished earthworks at both the Bristol and London end a quagmire, thus work was not completed until the end of 1839. The cutting is spanned by the main London to Bristol road, now the A4, with a three-arch brick bridge, while the bridge carrying a lesser road was of timber, and formed the forerunner of Brunel's many timber viaducts.

Liverpool shareholders, concerned that the railway to Bristol plus Brunel's ship the *Great Western* might take away some of Liverpool's trade, attacked Brunel. On 13 July 1838, George Gibbs, a member to the Bristol Committee of the GWR, wrote in his diary:

He [Saunders] showed me a letter he had just received from Brunel expressed in a cool and very proper way, but showing great feeling with regard to the loss of confidence which he believes he has seen on the part of the Directors and even of Saunders. Poor fellow, I pity him exceedingly, and I know not how he will get through the storm which awaits him. With all his talent he has shown himself deficient. I confess, in general arrangement; I mean in arranging his work in his own mind so as to enable him to proceed with it rapidly, economically and surely. There have been too many mistakes; too much doing and undoing. The draining, I fear, is imperfect, and the carriages made under his direction have not worked well; but I cannot help asking myself whether it is fair to decide on a work of this kind within a few weeks of its opening; and is not the present outcry created in a great measure by Brunel's enemies? I hear that at the meeting Brunel's dismissal is to be moved. Now the strong bias of my mind is that our only chance of comfort and safety is that our line should be carried out by Brunel with efficient assistance, and on a more stringent system of control, unless Stephenson will join him on the principle of abandoning his granite blocks and following out Brunel's wide gauge. It can only be done by Brunel himself, and, even if Stephenson would join him, I doubt much if they would work well together.

At the meeting on 16 July 1838 it was suggested that Robert Stephenson be brought in. Following this, on 20 July 1838 it was decided to invite Stephenson and James Walker to compile a report on the GWR. Stephenson was certainly a railway expert, but Walker, after early experience on drainage and docks, possessed less expertise, having been appointed engineer to the Leeds & Selby and Hull & Selby railways. On 30 July 1838, it was decided that Nicholas Wood, a close associate of Stephenson, should also be asked to report. In the event, Stephenson and Walker declined the invitation and were replaced by John Hawkshaw, engineer to the Manchester & Leeds Railway.

Hawkshaw's report, at a cost of £7 7s 0d for each of the thirty-four days, criticised the expensive track and concluded that the broad gauge was unnecessary. Time proved that his comments were correct. Although Hawkshaw stated the obvious faults, he failed to suggest how meaningful improvements could be made. Hawkshaw made the inaccurate statement that the locomotives were too heavy. Henry Gibbs, chairman of the GWR, said that the report was 'a very ill-natured production from beginning to end, the greater part of which might have been written without coming near the line'.

It is not always appreciated the part Brunel played in the contest between Bristol and Liverpool – the former an old, established port and the latter an up-and-coming one. On the railway front the GWR shareholders in the north of England, known as the 'Liverpool Party', claimed that only George Stephenson knew how to build a railway and that Brunel and his innovations were quite hopeless. They thought their suspicions confirmed when his track proved rough and his engines so feeble. On 15 August 1838, they pressed for an independent engineer to examine and report on the state of the railway.

In a postscript of letter to T. E. Harrison dated 3 September 1838, Brunel wrote:

> In reply to your enquiries the road is running really very fairly. We go regularly now at 30 to 35 miles per hour. I think the line smoother than any other I have been on [unfortunately he had not experienced Locke's smooth track on the London & Southampton Railway]. Our traffic is very large and our trains heavy. We took down 400 Rifles and brought back 400 Grenadiers, arms and all – could not be much under 90 tons last week at about 32 miles an hour And you could stand up in a carriage without holding the whole way.

It is a pity that Brunel did not adopt the aforementioned Joseph Locke's track as used on the London & Southampton Railway. This consisted of wooden sleepers which had cast-iron chairs to hold the rails, a much cheaper and simpler method – but it may have been its very simplicity which put Brunel off and encouraged him to think of something more elaborate. It was Locke's and not Brunel's system that became the British standard for over a century.

When the meeting of 15 August reopened on 10 October Brunel's opponents were defeated, Gibbs saying: 'The Liverpool men brought forward their points very feebly. Brunel defended himself from their charges with coolness and great effect.'

Nicholas Wood's report did not appear until 12 December 1838 and consisted of eighty-two closely printed pages with an even longer appendix. Wood was assisted by Dr Dionysius Lardner, the man who had those strange ideas about trains in Box Tunnel and several misguided calculations. Although the report was rambling and non-committal, his only definite criticism was made against the design of the piles, which Brunel had already admitted were unnecessary the previous July. It seemed that the meeting would appoint a consulting engineer such as Locke or Stephenson. Such action would have been tantamount to Brunel's resignation as he had always firmly stated that he would not accept divided responsibility.

The directors found the most disturbing finding in the report to be Lardner's experiments with their best locomotive, *North Star*. Although capable of hauling 82 tons at 33 mph, to achieve 37 mph the load had to be reduced drastically to 33 tons, and to reach a maximum of 41 mph there could be only 16 tons behind the tender. At this speed it consumed 2.76 pounds of coke per ton per mile. Lardner attributed this dropping off in performance to wind resistance – ironical considering he had forgotten about wind resistance in his Box Tunnel calculations! Regarding Brunel and his reaction to the report, Gibbs commented on 13 December 1838: 'He is perfectly convinced that a great fallacy pervades it, as may be shown and proved by experiment, he proposes to devote all his mind and energies to show this in the next three weeks.'

The orifice of the blast pipe was found to be much too small and choked the engine at high speed, while the blast pipe was wrongly placed relative to the chimney and the blast did not have the correct effect on the fire. When these faults were corrected, the *North Star* with a load of 43 tons, at an average speed of 38 mph start to stop, reduced its coke consumption to only 0.95 pounds per ton per mile.

Wood and Lardner ran an experimental train on 27 September 1838 which unfortunately collided with another; further to this mishap, on 27 October one of Dr Lardner's pupils was killed on the line.

Throughout the autumn, proposals were made that Joseph Locke, an ex-pupil of George Stephenson and engineer to the London & Southampton Railway, should replace Brunel in his role as engineer.

Following the publication of Wood and Lardner's report, the directors visited Brunel at 18 Duke Street to tell him of their decision to appoint a second engineer. Brunel averred that his methods were correct, refused the appointment of a second engineer and offered his resignation instead. On 14 December 1838, the GWR director George Gibbs wrote in his diary:

> Brunel ... in a very modest way said that the evidence which was accumulating against him appeared to be too great to be resisted without injury to the Company, and therefore he was prepared to give way. He had no vanity of any kind. If it were necessary to yield, he had no objection to it being said that he had been defeated, for he felt confident in the correctness of his views and was sure that he should have opportunities of proving it.

At a meeting on 9 January 1839, Brunel's opponents proposed that 'the reports of Messrs Wood and Hawkshaw contain sufficient evidence that the plans of construction pursued by Mr Brunel are injudicious, expensive, and ineffectual for their professed object, and therefore ought not to be proceeded with'. At the vote the notion of appointing a consulting engineer to assist Brunel was rejected: 6,145 votes of the Liverpool Party (representing the GWR's northern shareholders) for the motion and 7,790 against. Thus the Liverpool faction was defeated.

In November 1839 a flood was believed to have rendered the Maidenhead bridge unsafe. Brunel wrote to Saunders:

> On Sunday morning about 2 o'clock, Mr Hammond was called out of his bed by Low from Maidenhead with a message from Bell that the Maidenhead bridge was reported by Ld Orkney [?] as in a dangerous state and that the 6 o'clock train must not go over it. He immediately got in a post-chaise, went to Twyford, got an engine and went to the bridge. As well as he could by lanterns he examined everything – could find of course nothing wrong and nobody on the line knew of anything wrong. He then went for Bell who could only say that he heard of this but, it seems, never took the trouble to go and see for himself, as if anything had been wrong it might have been necessary to send

for me, but went quietly to bed – leaving it to chance, I suppose, whether Hammond might be at home and whether the train might tumble in the river or not – in fact taking no more trouble in the matter than to forward this cock-and-bull story on to Hd who was ten miles off while Bell was within a few hundred yards. I shall have my own quarrel to settle with Bell.

Brunel liked to be in complete charge of all aspects of the operation and was slighted by the fact that Daniel Gooch, the locomotive superintendent, was able to report straight to the directors and not through him. One example of his irritation with being undermined by Gooch was in a letter sent on 11 January 1840, when the locomotive situation on the GWR was very serious:

> The *Dog Star* driving wheel is gone – it is barely capable of moving and quite unfit for use with a train and the North Star is going.
>
> Have you any new ones – or what do you propose to do – you must provide for them *immediately*. We are today as badly off for engines as we have ever been.
>
> The new tender alarms me very much, it is much too weak and is twisting – at first I supposed it bad workmanship but on examining it I observed a radical defect in the construction – the whole weight is upon the inside frame without any transverse support to the under frame except at the two ends.
>
> The necessary consequence is thus [Brunel provided a sketch] and horizontally this.
>
> This must be remedied immediately I was not at all aware of their construction I mean I had not looked into it or I should certainly objected to it. *You should try and let me know where better to assist you.*

Brunel's progress report to the directors in February 1840 stated:

> Beyond Reading and up to Dudcot [*sic*] a distance of 17½ miles, the ballasting is completed with the exception of two short lengths, together about 2½ miles. The difficulty of procuring

ballast for this part has been very great; the ground purchased for this purpose being under water, and it being necessary to resort to dredging the river to obtain gravel. The laying of the permanent rails is in a forward state, a single line being laid for 15 miles, upon which all the materials for the second line are carried and distributed at all parts so that this work will proceed rapidly.

Beyond Dudcot [*sic*], great exertions have been made to complete the line for opening to a point near Faringdon simultaneously with the opening to Reading, and there can be no doubt that this might have been accomplished during May, probably even April, had the season permitted it. A few weeks will complete the earth-work, and preparations are making for ballasting. If no further delays should now occur from the indirect consequences of the late wet season, the opening to Faringdon may be calculated upon in June or the beginning of July.

The section from Reading to Steventon, 10 miles from Oxford, opened on 1 June 1840. Near Steventon station Brunel built a large, Tudor-style residence for the superintendent of the line and also provided office accommodation for the directors and a room for their board meetings. The line onwards to Faringdon Road opened 20 July 1840.

Brunel and his assistant William Gravatt did not always see eye to eye. On 23 July 1840, Brunel sent him a letter accusing him of betrayal. Initially he addressed him as 'My Dear Gravatt', but then conspicuously struck out 'My Dear'! Then, on 4 June 1841, he wrote to Gravatt: 'How could you leave me uninformed of the deplorable state of the bridge near the New Cut [Bristol]?' And again on 16 June, 'If you will retire without raising a question as to the cause, I will be silent.' Gravatt refused.

This led to a particularly lively episode at the Bristol & Exeter Railway Company's general meeting on 2 September 1841 – in fact the *Bristol Journal* went so far as to comment that it was 'conducted with greater excitement than we have witnessed at any respectable meeting'. Gravatt read a statement against

Brunel and a three-hour altercation ensued. Gravatt eventually resigned.

Contracts were let for building the Bridgwater to Taunton section of the Bristol & Exeter Railway in the spring of 1841. The only major work was Somerset Bridge, a mile south of Bridgwater, where the line crossed the River Parrett. Here Brunel designed a 100-foot-span masonry arch with a rise of only 12 feet – and therefore nearly twice as flat as his most criticised bridge at Maidenhead. In August 1843, when it had been used for over a year, Brunel reported to the directors:

> With regard to Somerset Bridge, although the Arch itself is still perfect, the movement of the foundations has continued, although almost imperceptibly, except by measurements taken at long intervals of time; and the centres have, in consequence, been kept in place. Under existing circumstances, it is sufficient that I should state in compliance with a Resolution of the Directors, measures are being adopted to enable us to remove these centres immediately, at the sacrifice of the present Arch.

Six months later the directors informed shareholders that 'a most substantial Bridge has been built over the River Parrett without the slightest interruption to the traffic'. Between the original abutments Brunel had substituted a timber arch which did duty until 1904, when it was replaced by a steel girder bridge.

Many of the places where Brunel stayed were unsuitable for his family to accompany him, but when engaged on the Bristol & Exeter Railway he and his family spent some time in Swiss Villa, Locking Road, Weston-super-Mare. Now demolished, for many years it was the home of Sir John Eardley-Wilmot, one of the prime movers of the erstwhile Brean Down Harbour scheme.

At Uphill, just south of Weston-super-Mare, Brunel erected what was the highest and widest single-span bridge in the country. Now Grade II listed, it has a span of 110 feet and is 60 feet above the line. It was built economically as it used less masonry than an ordinary bridge, since it was supported without abutments by the cutting sides. Also, as it was built before the cutting was excavated to its full dimensions, no expensive centring was required. It was known as Devil's Bridge, after 'Devil' Payne, a cantankerous landowner who possessed the land needed for the railway and who held out for a high price.

On 21 February 1840, George Stephenson wrote a letter to Brunel recommending John Brunton for employment.

> Mr Brunton who was employed under my Son on the London & Birmn. Raily. From the commencement to its completion as one of the sub Engineers later under Gooch [Thomas Langridge Gooch, Sir Daniel's brother] and myself on the Manchester & Leeds Raily. Until the opening of the portion under his charge, is in want of a situation as a Railway Engineer. I believe he gave every satisfaction in his departments on both the above lines, if you can find him a situation I shall feel obliged.

At this time Brunel did not give him a post.

Although in August 1839 the GWR directors believed that the line between Bristol and Bath would open in the spring of 1840, nature had other plans for the project, and the unusually wet winter caused a four-month delay. In February 1840, Brunel explained in his report:

> At the Bristol extremity the floods in the Avon have interfered with the supply of building materials; and at Bath and in its immediate neighbourhood the unprecedented situation of a state of flood in the river for a long period and till within the last few days has rendered it impossible to carry on the works of the Bridges or even the Station, the site of which has been flooded. Such a complete suspension of the works at some points and such delays at others have resulted from these and other causes indirectly consequent upon them

that certainly no less than four months' additional time will be required for the completion of some of these works, the whole of which would otherwise have been finished within a month or two of the present time, which must delay the opening to the end of the summer instead of the spring.

The works of the Station at Bristol, including the viaduct and offices, are rapidly advancing; but at Bath the causes I have referred to have prevented till within the last few days anything more than the commencement of the approaches.

Between these two extremities all the principal works – the Tunnelling, Cutting, and Embankments – are so far completed that, had the weather permitted it, the ballasting and permanent way would have been by this time in a very forward state. The excavation of the Tunnels is everywhere opened throughout, and the only work remaining to be done to them consists of the formation of the permanent drains and the finishing of detached parts of masonry, which in the general progress of the work had been injured or condemned, and the completion of one of the tunnel fronts. A few weeks will complete everything but the permanent rails, but many parts of the line, long since prepared, have not been in a state to allow of men or horses passing over them without destroying that portion of the forming which the rains had allowed to be completed, so that not more than 2½ miles of ballasting have been actually finished.

The 'tunnel front' mentioned in the penultimate sentence was the west portal of Bristol No. 2.

By the autumn of 1839, the work most in arrears on the Bath to Bristol section of the GWR was the skew bridge across the Avon, immediately west of Bath station. Tenders for the 500 tons of ironwork required had been invited in May 1839, but difficulties arose, and with a view to faster completion Brunel decided to construct it of laminated timber. This would come to be his only bridge of this type. J. C. Bourne in his *History of the Great Western Railway* describes it:

> The angle at which the Bridge crosses the River is so considerable that, although the space from quay to quay is only 80 feet, the space traversed by the railway is 164 feet. The bridge is of two arches, each of 80-ft span. Each arch is composed of six ribs placed about 5 feet apart and springing from the abutment and a central pier of masonry. Each rib is constructed of five horizontal layers of Memel timber held together by bolts and iron straps. The end or butt of each rib is enclosed in a shoe or socket of cast iron, resting with the intervention of a plate upon the springing stones, the shoes on the middle pier being common to the two ribs. The spandrels of the four externals ribs are filled up with an ornamental framework of cast iron supporting the parapets. The interior ribs are connected by cross struts and ties. The cornice and parapet are both of timber, the latter is framed in open work of a lozenge pattern. The abutments are flanked by plain turreted piers, and the tow-path is carried on an iron gallery beneath the western arch.

Laminated arches normally had a lifespan limited to twenty-five years because they were too flexible – frequent flexing causing the laminations to separate, with the subsequent ingress of water leading to decay. This laminated bridge however had a life of thirty-eight years, only exceeded in durability by Valentine's glued-laminated bow-string arch carrying the East Anglian Railway over the River Wissey.

The opening of the GWR from Twyford to Reading on 30 March 1840 brought the first of Brunel's one-sided stations into use. It was an idea devised by Brunel for use in a situation where the station was on just one side of a town and circumvented passengers having to cross a line to board a train. Non-stop trains could hence run clear of the platforms. Both the Up and Down platforms were arranged on the same side of the track with the Up platform being logically at

The famous photograph of Brunel taken by Robert Howlett in 1857 in front of the massive chains of the ocean-going steamship SS *Great Eastern*. Author's collection.

Left: Brunel watches the attempted launch of the *Great Eastern*, November 1857. Author's collection.

Below: Brunel standing by the funnel of the *Great Eastern* on 5 September 1859 only ten days before his death. Author's collection.

Above: *Fire Fly*. Author's collection.

Below: The Thames Tunnel. Courtesy of Jonathan Reeve.

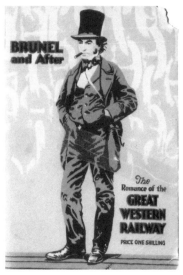

Above: Cover of book published by the Great Western Railway in 1925. Author's collection.

Above left: The SS *Great Britain* at the Gas Works Wharf, Bristol *circa* 1843. Author's collection.

Left: The forward drums for checking the launch of the *Great Eastern*, seen here November 1857. Author's collection.

A very early photograph of the completed Royal Albert Bridge with Saltash station in the foreground. Author's collection.

ove: River Wye Bridge, Chepstow. thor's collection.

ove right: River Wye Bridge, epstow. Author's collection.

ght: River Wye Bridge, Chepstow. thor's collection.

ryn Viaduct, *c.* 1910, Brunel also engineered structures in timber. Author's collection.

Above left: Royal Albert Bridge, east truss being raised in August/September 185 Author's collection.

Left: Royal Albert Bridge opening ceremony, 14 May 1859. Author's collection.

Below: The completed Royal Albert Bridge with steam locomotive crossin Author's collection.

ove: The stern of the *Great Eastern*. Author's collection.

low: The iron steamship SS *Great Britain*. Hand-coloured lithograph. Courtesy of
athan Reeve.

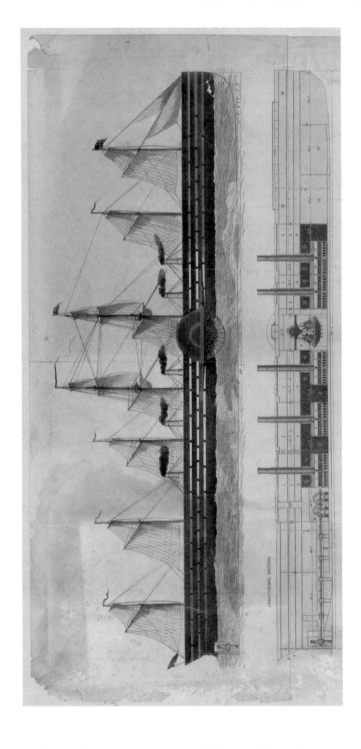

LONGITUDINAL SECTION

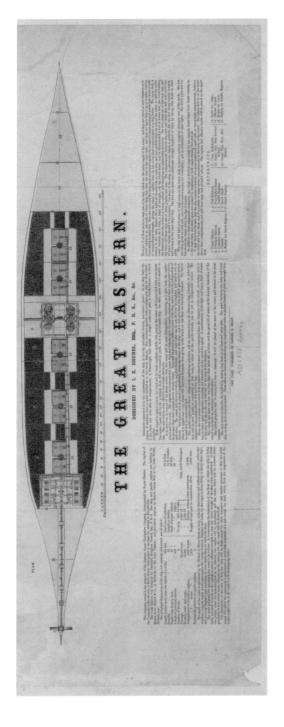

Above and opposite: Plan of the the SS *Great Eastern*. Courtesy of Jonathan Reeve.

Above: The steamship SS *Great Eastern*. Courtesy of Jonathan Reeve.

Below: Painting of the SS *Great Eastern* by Bristol-born artist Charles Knight. Hand-coloured lithograph. Courtesy of Jonathan Reeve.

above: Royal Albert Bridge under construction. Author's collection.

below: Royal Albert Bridge under construction. Author's collection.

Above: The Brunel family grave, Kensal Green Cemetery. Author's collection.

Below: The Great Western at Paddington station, the first ever railway hotel. Part of Brune revolutionary idea of integrated transport: from the opulent hotel guests could transfer direct their luxurious steam trains, race to Bristol, and board his ocean liner, the SS *Great Britain*, t largest and fastest ship in the world. Courtesy of Jonathan Reeve.

The Railway Station (1862) by W. P. Frith, painting of Paddington station as built by Brunel in 1854. A Fire fly locomotive can be seen top left. Courtesy of Jonathan Reeve b27fp2140.

Obverse of Brunel medal. Author's collection.

Reverse of Brunel medal. Author's collection.

Reverse of Thames Tunnel medal; tunnel opened 1843; opened to public 25 March 1843; *Great Britain* launched 19 July 1843. Author's collection.

Obverse of above medals. Author's collection.

Isambard Kingdom Brunel at his desk. Author's collection.

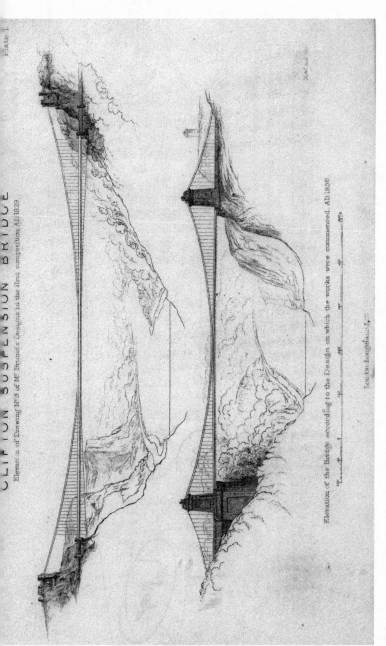

Elevation of Drawing Nº 3 of Mr Brunel's Designs in the first competition AD 1829

Elevation of the Bridge according to the Design on which the works were commenced. AD 1836

London Longman & Co

Clifton Suspension Bridge: top, Brunel's original design. bottom, Brunel's modified design dated 1836. (From *The Life of Isambard Kingdom Brunel Civil Engineer* by Isambard Brunel, published 1870.) Author's collection.

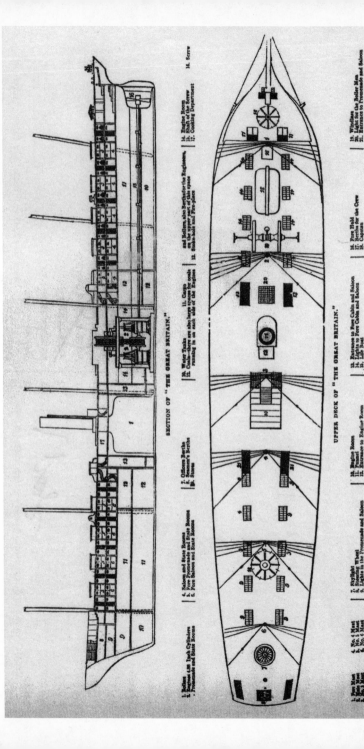

SECTION OF "THE GREAT BRITAIN."

1. Boilers
2. Engines 4.88 Inch Cylinders
3. Promenade and State Rooms

4. Saloon and State Rooms
5. Fore Promenade and State Rooms
6. Fore Saloon and State Rooms

7. Officers Berths
8. Seaman's Berths
9. Stores

10. Water Tanks
11. Coals—Slews are also large spaces for coals running to on each side of the Engines

11. Cargo
12. Coals—Slews are also large spaces for coals running to on each side of the Engines

and Boilers, also Berths for the Engineers, in the upper part of this space
13. Smoke-bank and Fire-place

14. Screw
15. Shaft of Screw
16. Cooking Department

UPPER DECK OF "THE GREAT BRITAIN."

1. Fore Mast
2. No. 2 Mast
3. No. 3 Mast

4. No. 4 Mast
5. No. 5 Mast
6. No. 6 Mast

7. Steering Wheel
8. Skylight
9. Lights in the Promenade and Saloon

10. Engine Room
11. Funnel
12. Entrance to Engine Room

13. Entrance to Fore Cabin and Saloon
14. Lights to Fore Cabin and Saloon
15. Life Boat

16. Fore Hold
17. Berths for the Crew
18. Capstan

19. Windlass
20. Light for the Boiler Men
21. Entrance to Promenade and Saloon

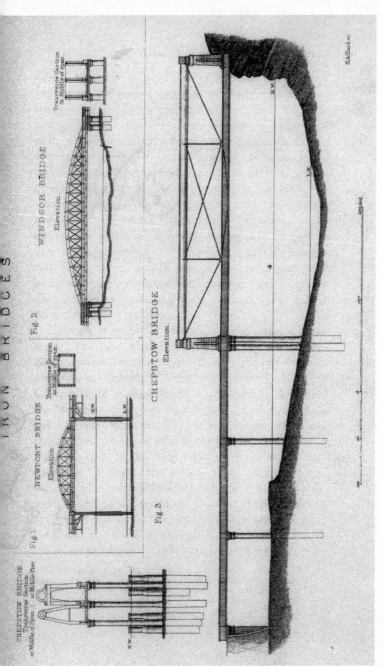

Elevation of Brunel's bridges at Newport, Windsor and Chepstow. (From *The Life of Isambard Kingdom Brunel Civil Engineer* by Isambard Brunel, published 1870.) Author's collection.

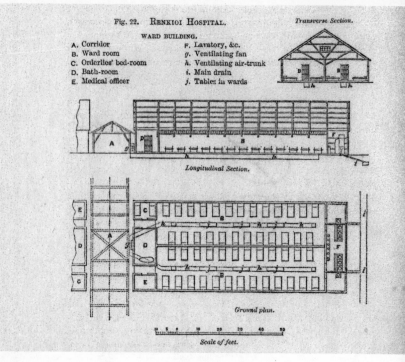

Fig. 22. RENKIOI HOSPITAL.

WARD BUILDING.

Transverse Section.

A. Corridor
B. Ward room
C. Orderlies' bed-room
D. Bath-room
E. Medical officer
F. Lavatory, &c.
g. Ventilating fan
h. Ventilating air-trunk
i. Main drain
j. Tables in wards

Longitudinal Section.

Ground plan.

Scale of feet.

The pre-fabricated Renkioi Hospital. (From *The Life of Isambard Kingdom Brunel Civil Engineer* by Isambard Brunel, published 1870.) Author's collection.

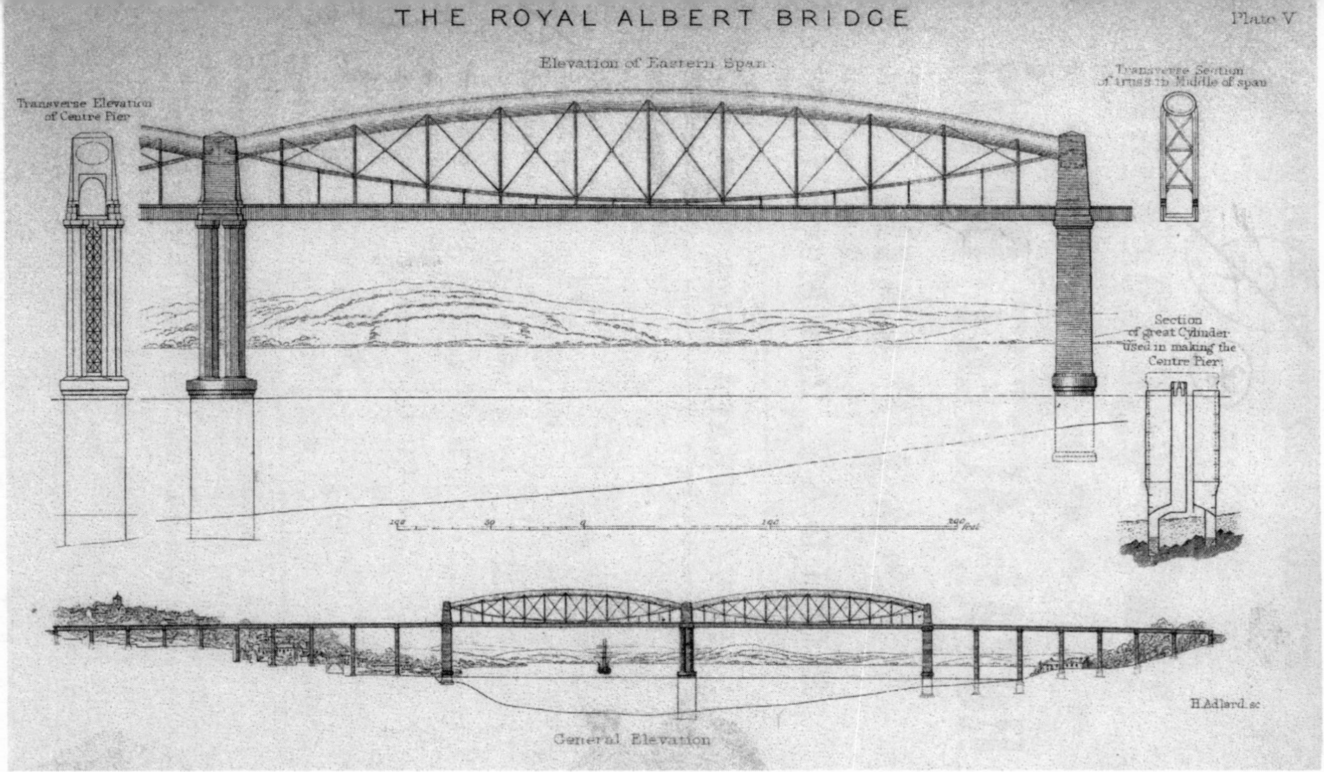
THE ROYAL ALBERT BRIDGE
Plate V
Elevation of Eastern Span.
Transverse Elevation of Centre Pier
Transverse Section of truss in Middle of span
Section of great Cylinder used in making the Centre Pier.
100
50
0
100
200 feet
H.Adlard sc.
General Elevation

The Royal Albert Bridge. (From *The Life of Isambard Kingdom Brunel Civil Engineer* by Isambard Brunel, published 1870.) Author's collection.

BRISTOL & EXETER RAILWAY.

VISIT

OF HIS ROYAL HIGHNESS

THE PRINCE CONSORT,

TO THE

OPENING

OF THE

ROYAL ALBERT BRIDGE,

AT

SALTASH,

ON

MONDAY, 2nd May, 1859.

ROYAL TRAIN TIME BILL.

DOWN.		DEP. A.M.	ARR. A.M.	UP.		DEP. P.M.	ARR. P.M.
WINDSOR		6 0		SALTASH		—	
Bristol			8 35	Cornwall Junction			—
"		8 45		"		6 50	
Taunton...			9 35	Newton			—
"		9 58		"			
Exeter			10 25	Exeter			8 15
"		10 35		"		8 25	
Newton			11 5	Taunton...			9 12
"		11 10		"		9 15	
Cornwall Junction			12 0	Bristol			10 5
"		12 5		"		10 15	
SALTASH			12 15	WINDSOR			12 50

The following arrangements will be necessary for the proper working of this Train, which must be strictly attended to:—

The 7.50 a.m. Down Passenger Train is to Shunt at Tiverton Junction.

The 8.0 a.m. Goods Train Down will not start from Bristol until after the Royal Train.

The 8.0 p.m. Up Train is to Shunt at Tiverton Junction.

The 9.20 p.m. Short Train from Weston is to Shunt at Yatton.

Bristol, 29th April, 1859.

A Bristol & Exeter Railway Royal Train timetable for the opening of the Royal Albert Bridge, 2 May 1859. Author's collection.

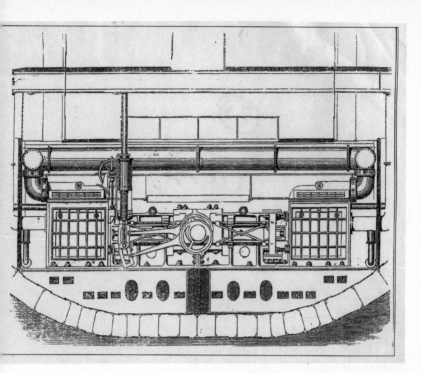

Above: The screw engines of the Great Eastern. Author's collection.

Right: GWR's first public timetable as issued in The Times, June 1838. Author's collection.

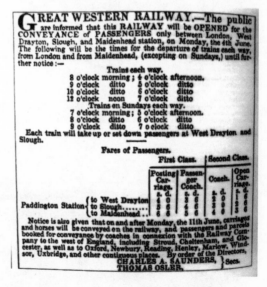

GREAT WESTERN RAILWAY.—The public are informed that this RAILWAY will be OPENED for the CONVEYANCE of PASSENGERS only between London, West Drayton, Slough, and Maidenhead station, on Monday, the 4th June. The following will be the times for the departure of trains each way, from London and from Maidenhead, (excepting on Sundays,) until further notice:—

Trains each way.

8 o'clock morning ; 4 o'clock afternoon.
9 o'clock ditto 5 o'clock ditto
10 o'clock ditto 6 o'clock ditto
12 o'clock noon 7 o'clock ditto

Trains on Sundays each way.

7 o'clock morning ; 5 o'clock afternoon.
8 o'clock ditto 6 o'clock ditto
9 o'clock ditto 7 o'clock ditto

Each train will take up or set down passengers at West Drayton and Slough.

Fares of Passengers.

Paddington Station	First Class.		Second Class.	
	Posting Carriage.	Passenger Coach.	Coach.	Open Carriage.
	s. d.	s. d.	s. d.	s. d.
to West Drayton	4 0	3 6	2 6	2 0
to Slough......	5 6	4 6	3 0	2 6
to Maidenhead..	6 6	5 6	4 6	3 6

Notice is also given that on and after Monday, the 11th June, carriages and horses will be conveyed on the railway, and passengers and parcels booked for conveyance by coaches in connexion with the Railway Company to the west of England, including Stroud, Cheltenham, and Glocester, as well as to Oxford, Newbury, Reading, Henley, Marlow, Windsor, Uxbridge, and other continuous places. By order of the Directors,

CHARLES A. SAUNDERS, } Secs.
THOMAS OSLER.

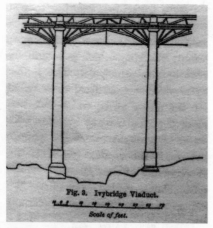

Fig. 3. Ivybridge Viaduct.

Scale of feet.

Above left: Robert Pearson Brereton. Author's collection.

Left: Ivybridge viaduct. Author's collection.

Below: An up train crosses Chepstow bridge. Author's collection.

Letter from Brunel to Michael Lane, a former engineer-in-chief of the GWR.

GWR's first public timetable for opening 4 June 1838. Author's collection.

Daniel Gooch. Author's collection.

the London end. Similar stations also appeared at Slough, Taunton, Exeter, Newton Abbot, Dawlish and Gloucester. Disadvantages were discovered, such as the number of cross-overs required, not to mention that when a station required additional platforms for future expansions, passengers still needed to cross the tracks.

The line between Bristol and Bath opened on 31 August 1840. In July 1839, Thomas Osler, secretary of the Bristol Division of the GWR, had written to his London counterpart regarding the architecture of Temple Meads central station. He observed that Brunel was instructed to prepare plans for offices

> only as were requisite and which were to be devoid of ornament as was consistent with decent sightlines. The first of the sketches exhibited a plain specimen of what I believe is now called the 'Tudor' style; the second consisted, I think of as thoroughly naked an assembly of walls and windows as could well be permitted to enclose any Union Poor House in the Country. A single glance at the two seemed to indicate that however agreeable the former might be to our tastes, the latter was the thing for our pockets, but when the Directors placed both Designs in the hands of a leading architect in the place they found, to their surprise, that the cost of the 'Tudor' front would exceed that of its Quaker companion by just 90 pounds.

The fact that trains at Temple Meads, the Bristol terminus, had to be on first-floor level in order to give sufficient waterway clearance, allowed Brunel to design the station like a Tudor mansion, thus giving a grand finale to the building style that characterised many of the bridges, tunnels and stations westwards from Bath.

This imposing station at Bristol, built on a field of the same name, was constructed on brick arches fifteen feet above ground level in order to give clearance to the waterway. On the north and old departure side on the ground floor were the booking offices. Luggage was carried up in lifts, but passengers used stairs to reach the platform. The walls were in Bath stone, while the impressive timber-cantilever train shed roof had a span of 72 feet, unsupported by any cross-tie or abutment, being carried on octagonal iron piers with four-centred arches. Each principal was formed of two frameworks like cranes, meeting in the centre of the roof, the weight being supported by the octagonal columns and the tail ends of the frames held down by the side walls. As the two frames did not press against each other where they met, there was no outward thrust. The octagonal piers were rather close to the platform edge and caused an awkward obstruction to passengers – Brunel should have spotted this problem.

The glazed roof had a length of 220 feet and the platforms extended for a further 200 feet. Beyond the train shed, very much like the nave of a church with the platforms as aisles, was the flat-ceilinged chancel for locomotive storage. Forsaking the Gothic, the ceiling was supported by closely spaced, slender, unembellished iron columns. Above and in front of this engine shed was the four-storey office block – and it was not unknown for steam and smoke from the engines to seep up through the floorboards.

The symmetrical Tudor façade facing Temple Gate had single turrets and a centre oriel above which are the arms of the cities of Bristol and London, these being adopted by the GWR for its own arms. There were two flanking gateways, the departure side on the left featuring a clock. The whole impression of the station was that of a gentleman's country seat, and it would have boosted the confidence of nervous early travellers.

To try to avoid earth slips, Brunel endeavoured not to make embankments during wet weather; nevertheless two near Wootton Bassett caused considerable difficulties, and in February 1840 subsequently he reported to his Board:

At Chippenham the works have proceeded well but in the neighbourhood of Wootton Bassett such have been the effects of the weather than it is probable that time and expense might have been ultimately saved by totally suspending the works during the last autumn and winter.

The quantity of work done during this period has been so limited that it would have required a few weeks only of summer weather to form the same extent of earthwork in a more substantial manner without incurring the same risk of future delays in the progress of the contracts from the slipping of earth excavated and thrown into embankment in a wet state. Arrangements are now making for redeeming as far as possible the time which has been lost, by prosecuting the works by the use of locomotive engines and other means with every possible vigour and despatch during the coming season.

Six months later Brunel resorted to a technique of side cutting to obtain material for wider embankments. As the material was immediately to hand, it 'has not only improved the quality of the earthwork, but has enabled me to expedite the final completion, and at the same time to perform the great bulk of the work during the Summer instead of the Winter months'.

Unfortunately the building of wider embankments failed to solve his problem of slipping earth, so in the spring of 1841, in order to complete the work, he tried driving piles down each side of the embankment into solid ground and further lashing the tops of the piles together with chains. Even this failed to solve the problem as the clay continued to move. A slip at the foot of Wootton Bassett incline on 7 September 1841 derailed the Down mail. The section had been only opened on 31 May 1841.

When engaged on this troublesome length of line he wrote a letter to his wife:

My Dearest Mary,

I have become quite a walker. I have walked today from Bathford Bridge to here – all but about one mile, which makes eighteen miles walking along the line – and I really am not very tired. I am, however, going to sleep here – if I had been half an hour earlier, I think I could not have withstood the temptation of coming up by the six ½ train, and returning by the morning goods train, just to see you; however, I will write you a long letter instead. It is a blowy evening, pouring with rain, my last two miles were wet. I arrived of course rather wet, and found the *Hotel*, which is the best of a set of deplorable public houses, full – and here I am at the 'Cow and Candlesnuffers' or some such sign – a large room or cave, for it seems open to the wind everywhere, old-fashioned, with a large chimney in one corner; but unfortunately it has one of these horrible little stoves, just nine inches across. I have piled a fire upon both hobs, but to little use, there are four doors and two windows. What's the use of the doors I can't conceive, for you might crawl under them if they happened to be locked, and they seem too crooked to open, the two ones with not a bad looking bit of glass between, seems particularly friendly [*sic*] disposed. [He drew here a cartoon of inward-leaning doors each side of a Georgian mirror.]

The window curtains very wisely are not drawn, as they would be blown right across the room and probably over the two extra greasy muttons [candles] which are on the table, giving just light enough to see the results of their evident attempt to outvie each other, trying which can make the biggest snuff. One of them is quite a splendid fellow, a sort of black colliflower[*sic*] and I don't like to destroy him, so I send you a picture of him. [Drawing of smoking candle.]

I hope this very interesting letter will reach you safely, dearest.

I believe, love, I must be at Bristol from Saturday till Wednesday or Thursday next. I will let you know more certain tomorrow, but answer me by return of post, and then I can arrange accordingly. I may probably come home for Friday, and then we return together. There is a horrible harp, upon which really and honestly somebody has every few minutes for the last *three hours* been strumming chords always the same.

Goodbye my dearest Love.

Perhaps it was just as well that he was not at home too much, for his brother-in-law John Horsley recalled that his snoring was 'prodigious'.

Early on 25 October 1840, Brunel witnessed his first railway accident. It was still dark as he was waiting on the platform at Faringdon Road station, then the temporary terminus, for an engine to take him to London. To his horror he saw a Down goods approaching unusually fast. Despite shouts from Brunel and others on the platform, coupled with the efforts of the guard in the open wagon next to the locomotive containing four third-class passengers – in the early days this class travelled by goods train – the train sped through the station and burst through the closed doors of the engine shed. The driver, seen motionless on the footplate of the 2-2-2 *Fire King*, was killed, and four passengers and the guard injured. It transpired that the driver was asleep, suggesting that he must have been on duty for many hours to doze on such an uncomfortable place as a footplate.

On 29 October 1840, Brunel presented Gooch, in his capacity as the Locomotive Engineer and Superintendent, with *Rules & Regulations to be Observed by GWR Enginemen & Firemen*. On the same day he wrote to Gooch: 'J. Hill brought up the *Cyclops* in 27 minutes from Slough [this would have required an average speed of 40 mph], following the short train into Paddington within three minutes. This work must be put a stop to effectually. The Directors have determined to fine him ten shillings.' The humane side of Brunel was shown by the fact that after a chat about safety concerns – brakes and signals were not very effective at that time – he gave him ten shillings from his own pocket.

Brunel sent one of his assistants, Fripp, a severe letter on 12 October 1840:

Fripp. Plain gentlemanly language seems to have no effect on you. I must try stronger language and stronger methods. You are a cursed, lazy, inattentive, apathetic vagabond and if you continue to neglect my instructions I shall send you about your business. I have frequently told you, amongst other absurd, untidy habits, that of making drawings on the backs of others is inconvenient. By your cursed neglect of that you have wasted more of my time than your whole life is worth. Looking for the altered drawings you were to make of several things at the station and which I have just found – they won't do. I must see you on Wednesday. Let me have no more of this provoking conduct or of the abominable and criminal laziness with which you suffer contractors to patch and scamp their work.

Six months later, on 21 April 1841, Fripp was sent another missive:

A long time ago I gave you instructions respecting certain details in the rooms at Bath station. Notwithstanding this I have to attend to every detail myself. I am heartily sick of employing you to do anything that, if I had ten minutes to spare, I would do myself. If you go over to Bath on Friday and can be ready with the outlines of the finished rooms you may still save me some trouble – if you wish to do so. If not *pray keep out of my way or I will certainly do you mischief* you have tried my patience so completely.

The GWR was extended to the 80¼ mile post at Hay Lane on 17 December 1840. On 13 September 1840, Gooch wrote to Brunel suggesting that the best site for a locomotive depot was at Swindon. Brunel concurred and designed the works buildings and 300 nearby houses for the workers.

Although Brunel was a great letter-writer, in 1841 he was taken to task by the Cheltenham & Great Western Union Railway:

The Directors also especially call Mr Brunel's attention to the importance of his answering the letters addressed to him by the Secretary under the orders of the Board, when circumstances may not allow of his keeping appointments made with them, which would at least prevent their being called from their homes unnecessarily and the disturbance of the appointments of many persons, whose engagements, Mr Brunel's own experience will have taught him, are of great importance to individuals.

More demands soon began to pile up against a beleaguered Brunel. Soon after the opening of the Cirencester to Swindon line on 31 May 1841, the embankment began to slip. Early in December a worried passenger wrote to the Board of Trade:

> I returned by the railway and as far as Swindon all was very well, notwithstanding the wet, but from Swindon to Cirencester I was horrified at seeing the road I was passing over, and nothing shall tempt me to do it again. One line of rails has slipped for a mile or two completely away and the trains travel on the other line, which appears just hanging by a thread, and this on a precipice of 40 to 50 feet.

Such a letter demanded immediate investigation, and the Board of Trade officer General Pasley reported:

> Fortunately the ground on the western side of the embankment remained firm, so that the slips took place on the eastern side only, where the clay, almost in a fluid state, gave way and moved towards the adjacent cutting, this movement taking place below the surface, as was proved by the remarkable fact that some very strong piles, which had been driven along the bottom of the embankment, were forced forward out of their original line, moving along with the clay; and in one part in particular some of them are now to be 70 feet in advance of their former position. This movement was described to me as having been very slow, so that if carefully watched, and men stationed to stop the trains, no danger can arise from it; but it was so powerful on the east side that the ground under the rails there sank no less than three feet in 24 hours in the worst part.
>
> To make good the embankment Mr Brunel has caused soil of a better quality to be brought from a hill at the north end of it, and to be continually laid on the east side, using the rails on that side for the transport of this earth; and having found piling to be of little use, he has directed a dry wall of rubble stone, 12 feet thick, to be built at the bottom of the slope to a depth of ten feet, which is equal to that of the cutting, as a retaining wall, to prevent

further movement at the base of his embankment towards the ditch or deep cutting on that side; which, as a further precaution he has ordered to be filled up opposite those places where the greatest movement of the moist clay took place. These measures will, no doubt, prove effectual, for, as I said before, the western line of rails is perfect throughout, and the eastern line is now only about 15 inches lower than the other in the worst part, and is being gradually brought up to its proper level, which Mr Brunel hopes to accomplish in four or five weeks.

Brunel's resident engineer, Charles Richardson, in his address to the Geological & Engineering Sections of the British Naturalists' Society in May 1891, gave more details:

To stop these slips, Brunel, after inspecting them carefully, made up his mind to drive a row of strong piles along the 'cess' as rapidly as possible. He sent down half a dozen pile engines with 30-cwt monkeys, and piles were obtained for beech trees in the woods in the neighbourhood of Chalford Valley. These were cut into piles not less than ten inches in diameter, and twenty-feet lengths. Everything was on the ground in three or four days. The 'cess' made a capital floor for the pile engines to stand upon; the piles went easily through this plastic clay, each engine being able to drive ten or a dozen piles in the day; and in less than a couple of days all the piles upon a 100-yards slip were driven in a row. But the slip took no notice of them, and the pile-heads, which stood up a foot or so above the ground, kept on advancing along with the 'cess', but with an increasing prominence of curvature in the centre.

Brunel, determined to keep the traffic going, arranged for train-loads of sand from Swindon to be brought between passenger trains, and cast out by a crowd of men to keep the working line constantly raised. But the slip went down still faster; so the line was raised by placing beds of faggots close together under the sleepers, and then tipping sand over the faggots to steady them.

Richardson continued:

I was on the bank night and day during the worst time, and saw the trains over. We sometimes lifted the rails quite three feet between two trains, packing the faggots under the train, and then tipping the sand among the faggots. It was a 'broad gauge' line with longitudinal sleepers, otherwise it could not have been kept safe for the trains, and I took particular care to have the rails in a straight line, well to gauge and level across.

We could not keep the road up to its proper level, and at the worst times it was eight or nine feet below the level in the middle of the slip; but I eased off the gradients at the ends to make them less steep, and in this condition the trains took passengers across for many days, until we succeeded in checking the progress of the slip by appliances at its foot. In the meantime, however, the descent the trains had to go down was fearful, and most of the engine-drivers went at it so timidly that I was afraid that they would never get up the rise again on the other side. There was one man, a very plucky driver, Jack Hurst, whom the other men nicknamed 'Hell-fire Jack'. He always looked for me when he came up to the slip, and when I gave the signal 'All right', he put on steam boldly and went over without trouble. He offered to stay there and take all the trains over, but was not allowed to do this.

After a high lift of the road, and a quantity of fresh faggots and sand had been put under the track to pack it up, these materials lay, of course, very hollow; and as the heavy engine passed over, the road sank down a couple of feet under the wheels of the engine, much like a wave of the sea, running along in front of the wheels, but there was never the least hitch or accident.

As a celebration of its many successes and progress, on 13 June 1842, Queen Victoria made her first journey on the GWR, travelling from Slough to Paddington. The locomotive, *Phlegethon*, had Gooch and Brunel on the footplate.

9

SS *GREAT BRITAIN*, THE SHIP THAT CHANGED THE WORLD

On the maritime front, the economic rivalry soon further intensified between Bristol and Liverpool. As soon as the idea of a steamship service from Bristol to New York was announced, the Transatlantic Steamship Company was formed at Liverpool. Although this enterprise quickly failed, it was replaced by the North American Royal Mail Steam Packet Company, soon known by the shorter title the Cunard Steamship Company. These ships used the shorter crossing from Liverpool to Halifax and Boston, and won the lucrative mail contract.

Back in October 1838, when Brunel was standing in the harbour at Bristol, a channel packet *Rainbow* called. She was then the largest iron steamer in the world. Impressed, he foresaw its possibilities and sent his associates Christopher Claxton and William Patterson to Antwerp and back on her. Their reports proving favourable, he recommended that the Great Western Steam Ship Company's next vessel, the SS *Great Britain,* should be of iron. Apart from the engines, Brunel had not been all that enthusiastic about the *Great Western,* but this new iron ship was to be his special pride. Patterson designed the hull; each set of plans made her larger and larger until she reached 3,400 tons. On 18 July 1839, construction commenced at Bristol. It must be remembered that, concurrent with Brunel's design of the *Great Western* and the *Great Britain*, the GWR and its associate railways were at a critical stage of construction.

Brunel was particularly interested in the interior structure of the hull, the engine and the screw. He conceived of the hull as being like an iron railway bridge; in fact Brunel had designed a huge box girder for the hull of the *Great Britain* before Robert Stephenson produced similar box-girder bridges at Conway and Menai. The ship's design, due to its length of 322 feet, was particularly crucial because a wave lifting her midship would cause the bow and stern to droop, or sag in the centre if she was lifted at both ends. She was trussed and braced with a double bottom and very strong.

In July 1839, the hull of *Great Britain* was laid. Then, in the following spring of 1840, the *Archimedes* arrived in Bristol, driven by a screw rather than paddles. Brunel was impressed. Work on the *Great Britain* was suspended while the *Archimedes* was borrowed for trials.

Screw propulsion had many advantages: the centre of gravity was lower than with high paddle wheels; there was less wind resistance; when a ship rolled the screw would remain in the water, whereas a paddle wheel could roll out of the water causing the engine to race and incur damage; a screw offered more space below for cargo and passengers. Brunel wrote: 'The wisdom, I may almost say the necessity of our adopting the improvement I now recommend is too strong, and I feel it is too well founded, for me to hesitate or shrink from the responsibility.' Building recommenced in the spring of 1841 and the hull was redesigned for screw propulsion. Brunel wrote to Guppy:

If all goes well we shall all gain credit, but *quod scriptum est manet* [what has been written will remain]; if the result disappoints anybody, my written report will be remembered by everybody, and I shall have to bear the storm. I feel more anxious about this than about most things I have had to do with. All mechanical difficulties must give way, must in fact be lost sight of in determining the most perfect form.

The engine, built by Maudslay, Sons & Field, was required to sit lower in the hull than for a paddler. Brunel wrote to Joshua Field on 21 May 1841:

> Remember when I urged you some years ago to consider the importance of a compact engine instead of a beam engine – you were rather disposed to think it more prudent to keep to your old form. You have thought better of it since and invented a very beautiful arrangement – to keep pace with these improving times and don't grudge a little trouble to obtain what *will* be sought after.

Brunel had provided the basic design for the engine, taken from a patent of his father with two pairs of cylinders inclined upward at sixty-degree angles to each other, producing 1,000 horsepower and driving an overhead crankshaft.

Great Britain's success or failure depended on the success of the screw. If successful, she would be the fastest ship across the Atlantic, but as it was an unproven concept, plenty of difficulties could arise. There were certainly problems to solve: how could the shaft at the stern be made leak proof; what size and shape should the blades be and how many?

In 1841 Sir Edward Parry, Controller of Steam Machinery for the Royal Navy, suggested that, as Brunel was using a screw propeller in the *Great Britain,* such a device should be tested by the Admiralty in order to discover any advantages. His rationale was that a screw below water level was much less likely to be struck by gunfire than paddle wheels above the surface.

Brunel was not impressed with his dealings with the Admiralty as their way of operation was far distant from his. Within a fortnight, as consultant to the Admiralty he promptly had drawn plans for the engines, but when he inquired of the whereabouts of the hull to put them in was told that work had not started. At an interview with Sir George Cockburn, the First Sea Lord, he was shown a model of an old three-decker with a large slice cut out of its stern.

A label was fixed to it: 'Mr Brunel's Mode of applying the Screw to Her Majesty's Ships.' Brunel disclaimed responsibility, and when Sir George left the room he scratched off the label with his penknife. In truth, the Surveyor to the Navy had mutilated the model.

In due course, the Admiralty informed Brunel that the *Acheron* could receive the engines. Brunel observed that its hull was quite unsuited due to the shape of its stern. The Admiralty continued to drag its feet; four months later Brunel resigned from the project. This stirred the Admiralty to produce the *Rattler*. One of the tests involved *Rattler* having a tug-of-war with the otherwise similar paddle sloop *Alecto*. When both vessels were at full steam ahead, the *Rattler* pulled *Alecto* at 2.8 knots. Screw propulsion was adopted by the Royal Navy.

Of the eight screws tested by the *Great Britain* committee, the fastest speeds were reached with a screw designed by Francis Smith. It featured four wide blades which tapered towards the shaft. In a letter to Thomas Guppy, probably penned in August 1843, Brunel wrote: 'We must stick to four arms altho' I feel pretty confident that with our pitch Six would be the right number.' Initially the four-bladed Smith screw was to be adopted, but almost at the last minute a six-bladed screw with more open space between them was used, this being Brunel's intervention.

Meanwhile, the company's shareholders were angry at the delays in construction, coupled with its rising cost. The estimate of £76,000 had risen to over £100,000, but she had no less than two new distinctions: the largest ship built and the first iron steamer to cross the North Atlantic.

In February 1841, Brunel reported to the GWR directors:

In the immediate area of Bath much still remains to be executed on one contract. The diversion of the Kennet & Avon Canal, in the progress of which very serious difficulties had occurred, caused principally by continued wet weather at a critical period of the works, requires most attention; the retaining wall, however, is nearly completed, and when the course of the canal is diverted, which will shortly be done, the construction of the railway itself at this point is a simple and easy operation. With the exception of this one point, the works are in a sufficiently forward state between Bath and Bathford. At Bathford the bridge across the Avon is much in arrears, but the necessary means have been and shall be adopted for securing its early completion.

From this point to the Box Tunnel, the works are in a forward state; the long embankment requires but a small additional quantity for its completion, and we are commencing to form the surface preparatory top ballasting.

The small tunnel in Middle Hill and the adjoining cuttings are nearly finished.

The works between Chippenham and the Box Tunnel, which have generally been considered as likely to be the latest, are now in such a state that by proper exertion, their completion within the time required may be ensured; this exertion shall not be wanting on my part.

The Box Tunnel itself will be completed and open throughout from the western face to the Shaft No 8, which has always been considered as the eastern extremity, during the next month; and if the whole tunnel cannot then be said to be finished, it is only because the eastern end, which is entirely in rock, has been extended a few yards in order to diminish the quantity of excavation required in the open cutting. The permanent way in the tunnel will shortly be commenced.

The remaining section of the GWR line hitherto unopened, that from Chippenham to Bath passing through Box Tunnel, opened 30 June 1841, initially running with only a single line through the tunnel section while the other road was being completed. For the first forty-eight hours Daniel Gooch personally acted as pilot man to every train, thus ensuring that no head-on collision could occur. Brunel was temporarily lodging in Bath, and Gooch recalled that he 'was very kind to me in sending me plenty of good food etc to keep my steam up. Box Tunnel had a very pretty effect for the first couple of days it was worked as a single line, due to the number of candles used by the men working on the unfinished line. It was a perfect illumination extending through the whole tunnel nearly two miles long.' The suggestion was made that the tunnel should remain illuminated to calm anxious passengers, but Brunel said that this would be a waste of money as the tunnel was no darker than the rest of the line at night.

Gooch recalled in his diary: 'The question of working through the Box Tunnel up a gradient of 1 in 100 was a source of much anxiety to Mr Brunel ... I cannot say I felt any anxiety; I had seen how well our engines took their loads up Wootton Bassett [also on a gradient of 1 in 100] without the help of a bank engine, and with the assistance of a bank engine at Box I felt we would have no difficulty.'

A leading Victorian artist, John Martin, who produced illustrations for Milton's *Paradise Lost,* was a friend of Brunel and on one occasion travelled on the footplate of an engine driven by him through Box Tunnel at a speed reportedly close to 90 mph. It is possible that the design of the west portal of Box Tunnel was based on Martin's Egyptian-style pictures. An alternative influence may have been Vanbrugh's east gate at Blenheim Palace. This east gate, apart from being a monumental arch, was additionally a water tower, and this combination of splendour and utility would have appealed to Brunel.

Brunel was a particularly creative man and enjoyed integrating his personal interests and curios into his designs as well as catering to their basic function. Box Tunnel is straight and the rising sun shines through on 6 and 7 April (for some time these dates were queried, but were confirmed by British Railways' engineers who made observations in 1988). If there was no refraction, the sun would shine through on Brunel's birthday, 9 April, but Bessel's Refraction Tables were not readily available in England in the 1830s. An effect of which Brunel was unaware for was that the earth's atmosphere causes a slight bend of the sun's rays, which enables the sun to be seen rising three minutes before it is actually there, geometrically speaking, and likewise it is seen setting three minutes after it has actually gone.

Another example of Brunel's clever detailing can be found at the 198-yard-long Middle Hill Tunnel, set half a mile west of Box Tunnel. This structure has portals flanked by pilasters decorated with fasces – traditionally a bundle of rods carried in front of a Roman magistrate. Just as those rods were carried before a dignitary, so for an Up train Middle Hill Tunnel comes before the significant Box Tunnel.

Brunel's meticulous nature left nothing to chance and he visited various works which supplied the GWR. Following a visit to Messrs Stothert, Slaughter & Co at Bristol, he wrote to the owners on 24 September 1840:

Passing through your shops on Wednesday last [23 September] I happened quite accidentally to observe that one of the cylinders then fixed in the new engine [a Fire Fly class 2-2-2] had a large flaw or defect in it and upon enquiring of a workman who stated himself to be the Foreman of the job I was told that it was proposed to patch it – indeed such must have been the intention or the cylinder would not have been in place.

The flaw was so large that it could not escape attention of the most careless observer it was such as to render the cylinder perfectly useless. I do not believe that a workman would have thought of using it in the most contemptible worst managed shop in England except with an avowedly fraudulent intention and yet I find such a thing in an engine in which you profess to put the perfection of workmanship and materials and upon the success of which (as being your second and which I consider a much fairer criterion than a first) so much of the character of your House depends – you cannot yourselves deny that I have no alternative but to suppose either your Foreman or those called Foremen are utterly neglectful of their duties that they actually do not inspect the work and never saw the cylinder or that they could succeed in defrauding the Company and concealing such work. In either case your workmen are spoilt, they have learnt that they may scamp their work and it will be some time before this most mischievous effect is remedied.

All confidence on my part is completely destroyed – more than this I must believe that the workmen who found they might proceed to fix such a piece of work will as a matter of course do many worse thing in those parts less exposed to view.

I have felt with the Directors every desire to encourage your establishment but I shall neglect my duty to the Company if I did not now advise them to withdraw the order last given or at least to attach such conditions to it as will secure them, as far as possible, against loss by the consequence of bad workmanship. I stated to your foreman that I should require the defective cylinder to be broken in my presence as a security against it being used again. My object is principally for your advantage to shew the men that such work will not be allowed.

Brunel was invited to become engineer to various branch lines from the GWR. The Bristol & Exeter Railway was established in 1836, Brunel delegating William Gravatt to carry out the survey. The Cheltenham & Great Western Union Railway, whose Act was the same year, ran from Swindon to Gloucester and Cheltenham, while the South Devon Railway established in 1844 extended the broad gauge to Plymouth, and the Cornwall and West Cornwall Railways took it even further westwards. The Wilts, Somerset & Weymouth Railway ran from Bath and Chippenham to Salisbury and Weymouth. The South Wales Railway secured its Act in 1845. Brunel did not consider the River Severn a barrier to this expansion and on 30 May 1854 wrote, regarding a planned bridge: 'I believe firmly that before fifty years are over there will be one (or a tunnel).' In the event he had to make do with the Bristol & South Wales Union Railway from Filton to New Passage with a ferry over to Portskewett.

The engineer for this latter line was Charles Richardson. That Brunel had a forgiving side is reminded by the fact that on 14 September 1853 he wrote to Richardson:

… as to the apparent want of energy and activity on your part in attending to the Company's works, contrasted with an alleged devotion to amusement and amongst other things to cricket, I don't know why you should be less of a slave to work than I am, or Mr Brereton, or any of my assistants in town. It would rather astonish anybody if Mr Bennett should be a frequenter of Lord's cricket ground or practice billiards in the day time, and I don't know why a man having the advantages of country air and very light work should indulge them. You must endeavour to remove any such grounds of observation.

Yet exactly five years later, on 14 September 1858, Brunel invited him to become resident engineer on the Bristol & South Wales Junction Railway:

I want a man acquainted with tunnelling and who will with a moderate amount of inspecting assistance look after the Tunnel with his *own eyes*, for I am beginning to be sick of Inspectors who see nothing, and resident engineers who reside at home. The country immediately north of Bristol I should think a delightful one to live in – beautiful country – good society near Bristol and Clifton etc. I can't vouch for any cricketing but I should think it highly probable.

Charles accepted the post and held it at the time of Brunel's death. Later, Richardson was to become engineer to the Severn Tunnel project.

It was at this time, on 24 March 1841, that his father, Marc, was knighted.

The Bristol & Exeter Railway opened between Bristol and Bridgwater on 14 June 1841. Work on bridges had been delayed by the unusually wet winter of 1839/40. Brunel used 62 pounds per yard bridge rail on longitudinal baulks, though the spacing of the cross transoms was adapted to 11 feet, returning to the usual 15 feet between Whiteball Tunnel and Exeter.

For the impecunious Cheltenham & Great Western, South Devon and Cornwall lines, Brunel designed timber bridges and viaducts, cheaper than iron structures, and which could be later replaced by something more durable when the companies' finances had improved. Some of his timber viaducts had a life of over fifty years. Brunel devised four types: (1) Class 1 was of continuous laminated beam construction; (2) Class 2 had queen post through trusses; (3) Class 3 had a queen post deck truss; (4) Class 4 had a king through truss, or a king post and 'A' frame and offered maximum clearance below a span.

The viaducts were built of Memel yellow pine which was both cheap and long lasting. They were economical to maintain as any of the timber pieces could be renewed simply by unbolting and replacing it, no total line occupation being necessary. After 1914 Memel timber was no longer available and the only suitable replacement was Oregon pine, a quite uneconomic option as it only had a life of about eight years instead of thirty or more. Nevertheless, two of Brunel's timber viaducts constructed in this way, on the former Vale of Neath Railway near Aberdare, were not demolished until 1947. Both stood 70 feet above the valley floor – the Gamlyn Viaduct 600 feet in length across the Cynon and the Dare Viaduct 450 feet across the Dare.

It was an industrious time for the whole family. Seldom have a father and son lectured to the same learned audience as did the two Brunels at a meeting of the British Association at Plymouth in August 1841. Marc spoke on the Thames Tunnel and Brunel on an invention of Professor Moseley for recording the rate of progress of steamers and locomotives.

That Brunel was also showing signs of strain is revealed by a letter he wrote to Saunders in the summer of 1842, apologising for his not attending a board meeting:

I cannot get out early in the mornings and this evening I feel that it would have been impossible for me to have been at Steventon tomorrow and doubt my attending even the arbitration. My state of health indeed renders me very anxious to get away entirely for a week or ten days or I see no prospect of my getting well.

With his father honoured with a Knighthood, Brunel's own reputation had begun to reach an international audience. On 28 August 1841, Brunel wrote to the Genoa Railway's solicitor, Edwin Gower:

> I have hitherto invariably declined any foreign engagements for which of course I have had many proposals partly on account of my engagements here and partly, I may say principally, because of the difficulties of satisfying myself of the perfect respectability of parties, without more trouble and enquiry than I could devote time for, and also the similar difficulty of obtaining any redress or of clearing myself should the direction change hands and the promoters adopt any course towards myself of in the conduct of their affairs which I disapproved of.

In December 1841, Brunel agreed that he would visit Italy, spending at least twenty-one days there. His stipulations were that he was to be considered the principal engineer, be paid ten guineas a day from the date of leaving London until his return, in addition to travelling expenses of £166 10s 6d.

The work to be done was in three parts: the first was the preliminary report on the scheme and was completed in December 1843; the second portion in the following year was the consultation with the Piedmont government for the acceptance of the report; while the third section in 1845 consisted of the detailed design work for the scheme.

Although Brunel made two personal visits to Italy, the majority of the work was carried out by his staff under Benjamin Herschel Babbage, who left England in March 1842. Brunel himself followed in April with his brother-in-law John Horsley. They passed several consecutive days and nights in the carriage. On 14 May 1842, Brunel wrote to the railway, saying: 'I am convinced that the line now under consideration may be constructed at a very moderate average expense compared with any of the English Railways.'

Brunel was most anxious that Babbage check drawings by careful measurement and that nothing should be sketched by just the eye; during the winter period details of local floods should be also be carefully noted. He wrote to Babbage on 14 November 1842: 'In case of violent floods it would be very desirable to have hourly or even more frequent notes to show the wave of the flood if there be any. I will send you two or three Massey's logs for measuring the velocity.'

On 20 November 1842 the highly nervous Lady Holland intended travelling from Paddington to Chippenham, where she was to stay on Lord Lansdowne's estate at Bowood. Brunel personally arranged for her to travel in his britzska placed on a flat wagon, and in order to reassure her held her hand throughout the journey.

When she wished to return, he wrote:

My Dear Madam,
You will be surprised, I almost hope so, that I should write to dissuade you from coming by the railway, or at least, to prepare you for anything but a quiet journey. Shortly after writing to Mr Allen, last evening, I received your Ladyship's note, and a few hours after, I was summoned to several points of the line where the floods and excessive rains have created impediments. The trains continue to run with perfect safety, but we have had many tedious and unpleasant delays, and if we should be disappointed in the hope of a fine night, which the evening now promises, I do not think the journey would be as your Ladyship would like. I am obliged to go to another part of the line to-night, but I will return to Chippenham tomorrow morning, and shall hold myself, and all other things, ready for you; but I confess it will relieve my mind from some anxiety for your Ladyship, although it may disappoint me to find that you do not intend to travel by the railway.

I hope your Ladyship will pardon this very hasty and, I fear, almost illegible, scrawl, but I am keeping a train while I write it.

Your Ladyship's devoted servant,

I.K. Brunel

Lady Holland returned by road early in December. Brunel was evidently gaining the confidence of both the aristocracy and establishment at home.

The Thames Tunnel was officially opened on 25 March 1843. On 26 July 1843, Queen Victoria and Prince Albert made a royal visit. At 1.00 p.m. the tunnel was closed to the public, but those in it and those invited by the directors were allowed to remain. At 3.30 p.m. the royal party arrived by water at the Tunnel Pier, Wapping. The Queen walked through the tunnel as far as the shaft on the Rotherhithe side, but did not ascend.

One of the stallholders lifted down from his shelves the whole stock of tunnel souvenir silk handkerchiefs and laid them on the ground to form a pathway for the Queen. *Punch* commented:

> The handkerchiefs being three and sixpenny goods, when raised from the ground after the transit of the royal feet, may be said to have got up amazingly, for they were sold during the rest of the day at half a guinea each; and there is no doubt that any person requiring one of the identical handkerchiefs that the queen walked over, may be supplied on the same reasonable terms for some considerable period.

The heat in the tunnel was very great and Her Majesty and consort appeared anxious to return. When about halfway up the shaft, the people below commenced singing 'God save the Queen', the 600 voices reverberating through the structure. On reaching the Tunnel Pier, Her Majesty was saluted by fifty coal porters who sent up a tremendous cheer. She smiled and bowed to them before embarking again on the royal barge.

A million visitors passed through the tunnel during its first fifteen weeks of operation. As the toll was only one penny, down-and-outs used it for sleeping accommodation. One such individual was disturbed during the night by a party of Frenchmen yelling: 'Le Tunnel vos de von only leetle ting in veech Londres bate Paris, and dat vos made by a Frenchman!'

As Brunel was busy with his work which required him to travel so much, it seems unlikely that he was able to spend much time with his family, though when he was at home he enjoyed playing with his children. He loved performing conjuring tricks, and on 3 April 1843, the occasion of his son Isambard's sixth birthday, he intended performing a trick of placing a half-sovereign in his mouth and then abstracting it from his ear. But on this occasion, while speaking, instead of being palmed, it slipped down his throat. He appeared to be unaffected by the incident, but two weeks later suffered coughing fits. On 18 April the eminent surgeon Sir Benjamin Brodie diagnosed the coin in Brunel's right bronchus and it was decided to perform a tracheotomy. This operation was to be undertaken using an instrument almost two feet long designed by Brunel and known as 'Brodie's forceps' – even though it was Brunel who sketched it. When these forceps were pushed through the incision in the wind pipe, Brunel found it impossible to breathe, so the attempt to remove the coin was abandoned. He consequently missed the opening of the Bristol & Exeter Railway to the temporary station at Beambridge west of Taunton.

His father Marc, residing with him at this time, realised that centrifugal force could be the answer. He sketched out a board pivoted between two uprights on which he could be strapped down and then swung rapidly head-over-heels. Its first trial on 13 May 1843 brought on a violent fit of coughing and choking and the experiment was stopped. When the coughing subsided Brunel signalled that another try be made. This time he was swung the coin was ejected.

That evening he wrote to his friend Captain Claxton in Bristol: 'At four ½, I was safely and comfortably delivered of my little coin; with hardly an effort it dropped out, as many another has, and I hope will, drop out of my fingers. I am perfectly well, and expect to be in Bristol by the end of the week.'

Brunel's mishap was reported in *The Times*. Such was the intrigue of this engineered cure that when the coin fell out the author Thomas Babington Macaulay ran to the Athenaeum Club shouting: 'It's out! It's out!' and everyone knew what he meant. The event was immortalised in the Reverend Richard H. Barham's *Ingoldsby Legends*:

> All conjuring's bad! They may get in a scrape
> Before they're aware, and, whatever its shape,
> They may find it no easy affair to escape.
> It's not everybody that comes off so well
> From 'leger de main' tricks as Mr Brunel.

On 12 March 1843, Brunel wrote again regarding the Italian railway:

> In going over these plains, you must wind about to seek easy levels. *Long* undulating country like this is just that in which the best engineers fail. But if you bear in mind that the lead becomes so great and everything must go to spoil, and embankments be formed from side cutting while the value of the land is excessive, earth-work thus costs at least *double*, and yet the plain is our only resource for *economy*.

Brunel delivered his report to the Genoa, Piedmont & Lombardy Railway in December 1843. He recommended it to be a freight line rather than one intended for passengers, and to be a double track, 4 foot 8½ inch gauge line. He envisaged that the steep gradients out of Genoa could use inclined planes operated by water power.

He made a personal visit for the first three weeks of January 1844, but a few weeks later was upset by the directors' criticism of his report. Peace was quickly made and on 10 June 1844 work began on the line from Genoa to Alessandria. Babbage had taken a post with Maria Antonia Railway at Florence and was unable to assist, so R. P. Brereton took his place in July, and Brunel and his wife went out in August and September. Some further unexpected conflict arose with the hosts, however, and on 18 November 1844, Brunel wrote to Count Pollan, the King of Sardinia's ambassador in London, formally withdrawing from the project: 'My assistant, a peculiarly energetic persevering young man writes to me declining to remain as feeling disheartened at the constant interference with every detail – and at the entire absence of confidence.' It was a temporary setback in Brunel's Continental venture.

THE LAUNCH OF SS *GREAT BRITAIN* AND MORE RAILWAYS, 1843–47

Brunel's projects closer to home continued apace. In the autumn of 1843 he visited the experimental length of atmospheric railway, less than 2 miles long, laid by the Dublin & Kingstown Railway. He made a return visit in November 1844. It is likely that he used these two trips to plan a railway from Rosslare, opposite a planned harbour at Fishguard, to Dublin. Brunel was appointed engineer to the Waterford, Wexford, Wicklow & Dublin Railway, engaging B. Gibbons as his assistant, but the potato famine of 1845 and 1846 caused the scheme to fall into abeyance. Work restarted in 1853 and the Cork to Youghal and Dublin to Wicklow opened in 1855.

The SS *Great Britain* was eventually launched on 19 July 1843, the sixth anniversary of the launching of the PS *Great Western*. Prince Albert travelled down from Paddington in the last coach of a four-coach train. The locomotive Fire Fly class 2-2-2 *Damon*, decorated with flags, was driven by the locomotive superintendent Daniel Gooch. Brunel accompanied him on the footplate for the occasion.

An observer for the *Bath & Cheltenham Gazette* in its issue for 2 August 1843 produced one of the earliest logs of a railway journey:

Paddington departure 7.03 a.m., Reading arrival 7.49½ a.m. where it took on water and coke; average speed 46 mph.

Reading departure 7.52, Swindon arrival 8.48½ where it took on water and coke; average speed 44 mph.

Swindon departure 8.52, arrival Bath 9.42 where the Prince was presented with an address from the City Council; average speed 33 mph.

Departure Bath 9.49, arrival Bristol Temple Meads 10.9½; average speed 34 mph.

From Temple Meads Prince Albert travelled via Clifton and Hotwells to the Great Western Steamship Company's dockyard, where he inspected the SS *Great Britain* – she was certainly impressive. A military officer estimated that 4,000 soldiers might be drilled on her decks; her length from figurehead to taffrail was 322 feet, 80 feet longer than the longest contemporary British battleship. Below the lower saloon were two decks, one for 1,200 tons of coal and the other for 1,000 tons of cargo. The bulkheads of the engine room were made of iron in order to contain any outbreak of fire.

The prince then joined a banquet for 520 guests. At 1.30 p.m. water was admitted to the dock, the caisson at the mouth of the dock removed, and at 3.15 p.m. the SS *Great Britain* floated majestically out into the harbour – as opposed to a dramatic launch when a vessel plunged into the water. A bottle of champagne was handed to His Royal Highness who deputed Mrs Miles (wife of one of the directors) to christen the vessel. Unfortunately the tug PS *Avon,* which was to tow the *Great Britain* out into the river, hauled the tow line too soon and when Mrs Miles swung the champagne bottle, she was out of range. Prince Albert grabbed a spare bottle and successfully flung it at the receding bows. The prince then returned to the station.

The return train left Temple Meads at 4.17 p.m., passed through Bath station at 4.32 and arrived at Swindon for 5.16 p.m., where it took on water and coke, the average speed being 46 mph. Continuing on its route, it departed Swindon at 5.20, arriving at Reading for 6.12 p.m., where it took on more water and coke, averaging a speed of 48 mph. And for its final leg, the return train left Reading at 6.15, with an arrival at Paddington for 6.56, an average speed of 53 mph.

The fitting out of the SS *Great Britain* took interminably long. She was not completed until the summer of 1845 – seven years after the decision had been made to build her. Apart from the decks and furnishing, the ship was entirely of iron. The six masts were of iron and ropes in the rigging were of iron wire rather than hemp; this wire, being thinner than hemp, offered less wind resistance. The propeller shaft, 130 feet in length and weighing thirty-six tons, was water-cooled.

The SS *Great Britain* certainly startled the maritime world. In an age of sail, a huge iron-hulled vessel with a propeller, rather than conventional paddle-wheels, was simply revolutionary. Not all observers were quite ready for the bold new design. An American magazine wrote: 'If there is anything objectionable in the construction or machinery of this noble ship, it is the mode of propelling her by the screw propeller, and we should not be surprised if it should be superseded by paddle wheels.'

After her fitting out at Bristol there was the additional difficulty of leaving, as the lock was too narrow. The *Bristol Mirror*, a paper which broadly opposed Brunel, likened it to a weasel in a granary grown too fat to escape! Although the dock company assented to the temporary removal of masonry to allow the ship to pass, it insisted that an Act of Parliament be obtained to authorise this alteration. With permission obtained and supervised by Brunel, at 6.30 a.m. on 11 December 1844, as she was being drawn through the lock by two tugs, she stuck. With great presence of mind she was drawn back by the third tug – due to her length exceeding that of the lock with both gates open, had she remained when the tide fell all water in the harbour would have drained out, leaving much of her unsupported and thus probably causing severe damage.

Brunel spent the rest of the day and night supervising the removal of more of the masonry in order to catch the last of the spring tides. Due to this curtailment, Brunel wrote to the directors of the South Wales Railway apologising for his absence the next day:

> We have had unexpected difficulty with the *Great Britain* this morning. She stuck in the lock. We *did* get her back. I have been hard at work all day altering the masonry of the lock. Tonight, our last tide, we have succeeded in getting her through but being dark we have been obliged to ground her outside, and I confess I cannot leave her till I see her afloat again and all clear of our difficulty here. I have as you will admit much at stake here and I am too anxious to leave her.

Queen Victoria and Prince Albert visited the SS *Great Britain* on 22 April 1844. Although she had a capacity of 360 passengers, she sailed for New York on 26 July 1845 with but forty-five. She made her first crossing in fourteen days and twenty-one hours, and returned in thirteen and a half days. The *Great Britain* was found to roll excessively, particularly in a calm when the sails could not help to stabilise her; this caused sea sickness which led to inevitable poor publicity.

On her return to Liverpool an extra two inches of iron were riveted to each propeller blade. She left again for New York on 27 September 1845 with 104 passengers. On 1 October she was struck by a heavy squall which felled her foremast; three of the six propeller arms snapped off and, as they were all on one side, made her highly unbalanced. This crossing took eighteen days. On her return voyage she only carried twenty-three passengers who were no doubt highly concerned as the propeller gradually disintegrated to leave just part of two arms. After twenty days she arrived at Liverpool under sail, her first season proving a complete technical and financial failure.

The only solution, with substantial investment already made, was to spend more money on the *Great Britain*. It was hoped to correct the rolling problem by adding two bilge keels projecting two feet outwards. The iron rigging had proved unsatisfactory and was replaced with ordinary hemp, while the new propeller was a strong piece of cast iron with four blades. Joshua Field improved the engine and boilers further to offer greater power and speed.

Only a few letters from Brunel to his wife Mary survive, one being written at Taunton on 17 April 1844 on the occasion when his young son Isambard Brunel Junior, at the age of seven, was being sent to boarding school:

> My Dear Love,
> Here I am on my way to Exeter. I have every reason to believe that, although I may be at home on Thursday, I shall be away again on Friday.
> I hope, dearest, you are feeling well and happy. You are wrong in supposing that I cannot feel your parting from dear Isambard. I hope the poor little fellow is not very unhappy but it is what all must go through and he has infinitely less cause for pain than most boys in beginning. I made my beginnings in ten times worse circumstances and now he will soon get over it. Give my love to the dear boy and tell him I have smoked his cigar case twice over. Adieu, dearest love to Baby,
> Yours devotedly,
> I.K. Brunel

Brunel's son Isambard shuffled rather than walked. A simple operation could have cured it, but contemporary thought was that surgery should be avoided if possible and Mary always protested, 'I never would let the surgeon's knife touch him.'

Brunel did indeed provide his son with good 'beginnings'. Young Isambard joined Harrow School in 1852, went up to Oxford in 1856, gained a BA in 1860 and an MA in 1863, and was called to the Bar the same year. He became Doctor of Law in 1870 and was Chancellor of the Diocese of Ely from 1871 to 1893.

In 1846, the SS *Great Britain* made two more round trips in acceptable time, but experienced several more mechanical mishaps. Then, starting her third voyage on 22 September 1846, she ran aground in Dundrum Bay, on the north-east coast of Ireland. Brunel was too busy on other projects, particularly the South Devon Railway, to visit her immediately, but was able to go that December and reported:

I was grieved to see this fine ship lying unprotected, deserted and abandoned by all those who ought to know her value, and ought to have protected her, instead of being humbugged by schemers and underwriters. The result is that the finest ship in the world, in excellent condition, such that £4,000 or £5,000 would repair all the damage done, has been left, and is lying, like a useless saucepan kicking about on the most exposed shore that you can imagine, with no more effort or skill applied to protect the property than the said saucepan would have received on the beach at Brighton. Does the ship belong to the Company? For protection, if not removal, is the Company free to act without the underwriters? If we are in this position, and if we have ordinary luck from storms for the next three weeks, I have little or no anxiety about the ship, but if the Company is not free to act as they like in protecting her, and in preventing our property being thrown away by trusting to schemers, then please write off immediately to Hosken to stop his proceeding with my plans.

As to the state of the ship, she is as straight and as sound as she ever was, as a whole. I told you that Hosken's drawing was a proof, to my eye, that the ship was not broken: the first glimpse of her satisfied me that all the part above her 5 or 6 feet water line is as true as ever. It is beautiful to look at, and really how she can be talked of in the way that she has been, even by you, I cannot understand. It is positively cruel; it would be like taking away the character of a young woman without any grounds whatever.

The ship is perfect, except that at one part the bottom is much bruised and knocked in holes in several places. But even within three feet of the damaged part there is no strain or injury whatever. There is some slight damage to the stern, not otherwise important than as pointing out the necessity for some precautions if she is to be saved. I say 'if', for really when I saw a vessel in perfect condition left to the tender mercies of an awfully exposed shore for weeks, while a parcel of quacks are amusing you with schemes for getting her off, she in the meantime being left to go to pieces, I could hardly help feeling as if her own parents and guardians meant her to die there. What are we doing? What are we wasting precious time about? The steed is being quietly stolen while we are discussing the relative merits of a Bramah or Chubb's lock to be put on at some future time! It is really shocking!

I should stack a mass of large strong fagots [sic] lashed together, skewered together with iron rods, weighted down with iron, sandbags, etc., wrapping the whole round with chains, just like a huge poultice under her stern and halfway up her length on the sea side. I have ordered the fagots to be begun delivering. I went myself with Hosken to Lord Roden's agent about it, and I hope they are already beginning to deliver them.

Claxton reported to Brunel that the sea washed the faggots away and received the following reply revealing some of Brunel's philosophy:

You have failed, I think, in sinking and keeping down the fagots [sic] from that which causes nine-tenths of all failures in this world, from not doing quite enough. I would only impress upon you one principle of action which I have always found very successful, which is to stick obstinately to one plan (until I believe it wrong), and to devote all my scheming to that one plan and, on the same principle, to stick to one method and push that to the utmost limits before I allow myself to wander into others; in fact, to use a simile, to stick to the one point of attack, however defended, and if the force first brought up is not sufficient, to bring ten times as much; but never to try to back upon another in the hope of finding it easier. So with the fagots – if a six-bundle fagot won't reach out of water, try a twenty-bundle one; if hundredweights won't keep it down, try tons.

In due course he succeeded in constructing a mattress which broke the force of the waves dramatically.

By the late spring of 1847 the leaks in her bottom had been sufficiently repaired so that they did not overcome the pumps, but the tides were insufficiently high to lift her over the reef. At last on 27 August 1847, she cleared the rocks by just 5 inches. It was planned to tow her to Liverpool the next day, but insufficient labourers appeared to man the pumps, so the idea was abandoned and she was grounded in Belfast Lough. A more reliable pumping gang was recruited at Belfast and she was towed to Liverpool the next day.

The *Great Britain* was saved, but the company ruined financially. She had cost over £200,000 but was sold in 1849 for just £25,000. Whilst the project proved Brunel's wisdom in building an iron ship for its strength, it was a wholly uneconomical endeavour during its time. She was refitted and worked other routes, mostly under sail.

Another professional failure of Brunel was the aforementioned Somerset Bridge, a mile south of Bridgwater, where the line crossed the River Parrett. Because at this location the railway was on a low embankment and he did not wish to have a rise in his level track, he designed a 100-foot-span masonry arch with a rise of only 12 feet, being as a result nearly twice as flat as his most criticised bridge at Maidenhead. In August 1843 when it had been carrying trains for over a year, he reported to the directors:

With regard to Somerset Bridge, although the Arch itself is still perfect, the movement of the foundations has continued, although almost imperceptibly, except by measurements taken at long intervals of time; and the centres have, in consequence, been kept in place. Under existing circumstances, it is sufficient that I should state in compliance with a Resolution of the Directors, measures are being adopted to enable us to remove these centres immediately, at the sacrifice of the present Arch.

Six months later the directors informed shareholders that 'a most substantial Bridge has been built over the River Parrett without the slightest interruption to the traffic'. Between the original abutments Brunel had substituted a timber arch which did duty until 1904, when it was replaced by a steel-girder bridge.

The opening of the broad gauge Bristol & Gloucester Railway on 8 July 1844 meant that the broad gauge met at Gloucester with the standard gauge Birmingham & Gloucester Railway. The break of gauge was a particularly serious complication in the case of freighted livestock and it was claimed that the change from one wagon to another deteriorated the quality of the meat 'very greatly', and it was then found impossible 'to compel the animals taken from one carriage to enter another until an interval of repose in a field or stable had allayed their tremor and alarm'.

Brunel blithely stated that 'a very simple arrangement may effect the transfer of the entire load of goods, from the Waggon of one Company to that of the other' and that passengers 'will merely step from one carriage into the other at the same platform'. His 'simple arrangement' never appeared and the matter of transfers did prove a great inconvenience.

By 1844, Brunel was feeling the stress of being engaged to carry out so many simultaneous projects and, anticipating the market speculation of the Railway Mania, in September wrote to the mathematician and inventor Charles Babbage: 'Things are in an unhealthy state of fever here, which must end in a reaction; there are railway projects fully equal to £100,000,000 of capital for next year, and all the world is mad. Some will no doubt have cause to be so before the winter is over.'

The following year he wrote to his friend Adolph d'Eichthal:

The whole world is railway mad. I am really sick of hearing proposals made. I wish it were an end. I prefer engineering very much to projecting, of which I keep as clear as I can. The dreadful scramble in which I am obliged to get through my business is by no means a good sample of the way in which work ought to be done ... I wish I could suggest a plan that would greatly diminish the number of projects. It would suit my interests and those of my clients perfectly if all railways were stopped for several years to come.

From 1841, Brunel had been busy working on the Hungerford Suspension Bridge. On 1 May 1845 it opened, giving pedestrians and barrows from the South Bank easy access to Hungerford Market, situated at the west end of the Strand. Only 14 feet wide, it consisted of two side spans each of 343 feet and a central span of 676 feet in length. The tops of the two piers based on the riverbed were designed like an Italian campanile. As at Clifton, the chains ran over rollers so that the towers only had to bear a vertical and not horizontal load. On the opening day, from noon till dusk no less than 25,000 foot passengers paid their halfpenny toll. The footbridge was eventually demolished in 1862 to give room for the railway bridge into Charing Cross station, and its piers were used for the railway structure while the chains recycled for use in Brunel's bridge at Clifton.

In May 1845, Brunel planned a floating pier at Portbury near the mouth of the Bristol Avon, and to connect it with the city by an atmospherically worked railway. The Portbury Pier & Railway obtained its Act the following year, but was wound up in 1851 through lack of funds.

In July 1845, when a Royal Commission was investigating the gauge question, Brunel was asked why he had not used the 7 foot gauge on foreign lines where he had acted as engineer. He replied:

I did not think that either the quantities or the speeds likely to be demanded for many years to come in that country [Italy] required the same principles to be carried out that I thought was required here, and I thought it very important that they should secure the goodwill of certain other interests which would lead into or out of the railway, and as a question of policy as much as of engineering I advised them to adopt the gauge [1.43 metres]. I thought it was wise to conciliate the interest of the Milan & Venice Railway and others which are likely to be connected with us.

The Taff Vale Railway, incorporated in 1836, chose Brunel as its engineer. In this capacity he advised the company to use the standard gauge of 4 foot 8½ inches, partly due to the curvature required along the banks of the River Taff. This meant that in due course the broad gauge South Wales Railway attracted little mineral traffic into England, most coal travelling by standard gauge to ships at Newport, Cardiff or Llanelly. In fact, the South Wales Railway had no direct branches into the mineral and coal rich valleys.

Brunel always preferred actions to words and so suggested a practical test to determine the best gauge. The courses were Paddington to Didcot, 53 miles for the broad gauge, and Darlington to York, 44 miles for the standard gauge. At the locomotive trials both parties pre-heated the feed water in the tender but Bidder, Gooch's opposite number, had two engines in the trial, not one, *Engine A* and *Stephenson*. Following a disappointing run of *Engine A*, all hopes were pinned on *Stephenson*, but she derailed and thus proved that the broad gauge was the best. The broad gauge 2-2-2 Fire Fly class *Ixion* attained a maximum speed of 60 mph with an 80-ton train, while its rival could only manage 53¾ mph with 50 tons. With this result in mind, Brunel was bitterly disappointed that the Gauge Commissioners, although acknowledging the broad gauge's advantage for speed, recommended the adoption of the narrow gauge as the standard gauge of the country and that existing broad gauge lines be narrowed accordingly. The Commissioners realised that it was easier and cheaper to narrow the gauge of a railway than to widen it; the broad gauge's extant lines totalled only 274 miles, whereas the narrow [standard] gauge had 1,901 miles already laid.

In Brunel's sketch book was a drawing dated 10 July 1845 for a hydraulic hoist for transferring a 1-ton container which could hold 4½ tons. This useful device was never made and all transhipment between the gauges was carried out by hand. Locomotive superintendent Daniel Gooch revealed in his diary:

> It was a hard fight and there is no doubt, so far as the evil of the brake [*sic*] of gauge was concerned, we had a weak case altho' everything possible was done to strengthen it. A machine was constructed at Paddington by which loads could be easily transferred from one gauge to another by changing the body of [the] waggon, load and all, or by lifting the narrow-gauge waggons complete upon a broad gauge platform. We also schemed a waggon with telescopic axles, adapted to run on either gauge, but I never had any faith in any of these plans as workable in practice. A break of gauge meant the unloading one set of waggons & putting the goods into another, as has since been done in practice with a platform between two lines of waggons.

The GWR's secretary, Charles Saunders, replied to the Gauge Commissioners' report by making twenty-four points in favour, though as the Great Western Board and Brunel probably accepted the case for the broad gauge as lost, Brunel did not prepare a workable device to ease transfer at a break of gauge.

The gauge issue proved vexatious to more than just the company and its engineers. One passenger, disgruntled at the change of gauge, wrote to the GWR complaining that he had been 'obliged to exchange the company of a pretty young lady for that of an ugly old one with her parrot and pet cat'.

Perhaps the first line to defeat the broad gauge standard was the Oxford, Worcester & Wolverhampton Railway, nicknamed, not without cause, 'the old Worse & Worse'. Brunel had surveyed it as a broad gauge line and its construction was partly financed by the GWR, but because it also linked with the Midland Railway and the London & North Western Railway it was agreed that it should be of mixed gauge.

The broad gauge South Wales Railway, with Brunel as its engineer, received its Act on 4 August 1845. Brunel had originally planned that access from London should be via the GWR and the Cheltenham & Great Western Union Railway to Standish, followed by a mile-long bridge over the Severn from Frampton-on-Severn to Awre, but as the Admiralty disapproved of a blocking of the waterway, a detour had to be made via Gloucester, thus denying Brunel the opportunity of building another memorable bridge. He was, however, able to use his ingenuity to design a swing bridge at Over. Curiously, the South Wales Railway's Act required an opening bridge at Over, just west of Gloucester, despite the fact that just downstream was a stone-built road bridge which barred the way to any high vessels. Brunel's design was a skew bridge of 125 feet wrought-iron girders swinging on a central pier to provide two 50-foot waterways.

Between the end of 1845 and the summer of 1848, Brunel worked for the Maria Antonia Railway at Florence, the railway being somewhat unusually named after a Hapsburg princess. Brunel's former assistant B. H. Babbage was already employed by the concern and so came under Brunel's direction once again. As with the building of the Great Western Railway, Brunel left little decision to his underlings and from June 1846 till December 1847 sent guidance on the preservation of timber, station design, dimensions of bridges and quantities and qualities of rails. When the line opened early in 1848 Brunel gave operational advice. Subsequently the railway was reluctant to pay him, so on 13 June 1848 he wrote to the Italian engineer Bonfil:

My Dear Bonfil,
Are you aware that your Company owes me £2150 for moneys advance in salaries to Babbage etc. and for my own? It has been a most shocking piece of carelessness on my part to allow such an arrear accumulating but in the midst of a pressure of business it has occurred and I appeal to you individually to get this debt discharged.

Discussions dragged on and the matter was settled in August 1848 when Brunel was reluctantly paid £1,028 12s 10d.

On 5 May 1846, Brunel recorded in his diary:

I have just returned from spending an evening with Robert Stephenson. It is very delightful in the midst of our incessant personal professional contests, carried to the extreme limits of opposition, to meet thus on a perfectly friendly footing and discuss engineering points as if we were united. Stephenson is decidedly the only man in the profession that I feel disposed to meet as my equal, or superior, perhaps, on such subjects. He has a truly mechanical head and it is singular that we never differ in the end when we thus meet, although always differing apparently as black from white in our public discussion. I cannot help recording the great pleasure I derive from these occasional though rare meetings.

The Vale of Neath Railway, with Brunel as engineer, was authorised by an Act of 3 August 1846. Its secretary, Frederick George Saunders, was the nephew of Brunel's friend Charles Saunders. This line required significant engineering works, including substantial tunnels, deep cuttings and timber viaducts. Two of these latter, on the branch from Gelli Tarw to Cwmanan, lasted until 1947. Both stood 70 feet above the valley floor – the Gamlyn Viaduct 600 feet in length across the Cynon and the Dare Viaduct 450 feet across the Dare.

The Vale of Neath Railway was utilised for bringing large quantities of coal to the Swansea and Briton Ferry Docks. As South Wales coal was friable, in order to avoid breakage by tipping it down a chute Brunel introduced wagons carrying four iron boxes, each containing 2 ½ tons of coal. Machinery at the docks lowered the boxes into a ship's hold, the bottom of a box being opened and the coal falling out relatively gently.

The Cornwall Railway Act was also passed on 3 August 1846, which continued the South Devon Railway from Plymouth to Falmouth. The topography was similarly difficult; no less than thirty-four of Brunel's timber viaducts were required in the 53 miles between Plymouth and Truro. As the company faced financial strain following the post-Railway Mania depression, single track was laid, though earthworks, bridges and tunnels were made double broad gauge width.

Its major engineering feature was the high-level Royal Albert Bridge across the River Tamar. The river at Saltash has a width of approximately 1,100 feet. Two problems here faced Brunel: firstly, the Cornwall Railway was poor and so especial economy was required, and secondly, no central rock was available.

The original scheme was to have one span of 255 feet and six of 105 feet but the Admiralty required the design to be modified, as at Frampton-on-Severn, to offer greater side clearance and further demanded 100 feet of mast room at high water. Brunel's final design was a bridge with two main spans of 455 feet and seventeen approach spans, varying from 69.5 feet to 93 feet, supported on masonry piers. Its overall length was 2,200 feet.

Work on investigating a site for the central pier was undertaken by divers early in 1847. In 1848 more detailed exploration took place. To assist this, a wrought-iron cylinder 6 feet in diameter and 85 feet in length was floated into position and lowered into the riverbed through 16 feet of mud and clay to the solid rock beneath. Water was pumped out and the mud removed – the experience of working under water which Brunel had gained when making the Thames Tunnel proving useful. The cylinder was raised and lowered for 175 borings to cover the whole area to be used by the pier. As a final experiment, Brunel had a small masonry column built on the rock up to the level of the riverbed. Due to a lack of finance the project then lay in abeyance for three years.

The South Devon Railway between Exeter and Plymouth posed Brunel a new problem. Originally he had intended making a flat line along the coast to Torquay and then across the River Dart and through the South Hams district, but this would have involved expensive bridges and viaducts. However, because locomotives had become so much better at hill-climbing in such a relatively short period, he then decided to take a more direct route. This involved climbing for 2½ miles at 1 in 36/98, followed by a descent to Totnes, followed by a final rise of 9 miles at 1 in 46/90.

When Brunel was called before the Parliamentary Select Committee on 4 April 1845 to justify the atmospheric railway he replied rather vaguely:

> The whole system of working a line with the atmospheric system required a great deal of consideration and requires many new contrivances and I do not think I have at all completed, to my own satisfaction, all the details ... I think I see my way clearly to effect them but I should still hope to effect many improvements.

Following the passing of the railway's Act, the Messrs Samuda contacted the directors. They had patented an atmospheric system whereby instead of using locomotive power a pipe ran between the running rails. Into this pipe a piston slung below a special carriage was fitted. When air was withdrawn from the pipe ahead of the piston by means of stationary engines set at intervals along the railway, atmospheric pressure from behind would propel the special carriage and likewise the train to which it was attached.

In September 1844, Brunel and Gooch visited the atmospheric railway between Kingstown and Dalkey and were most impressed; trains here could climb gradients at 30 mph, a feat impossible with contemporary steam locomotives. They would have heard that Frank Ebrington had sat in the motive carriage when suddenly it was whisked away without its train. He covered the 1¾ miles between the termini in only one and a quarter minutes, making an average speed of 84 mph – probably the fastest man had ever travelled up to that date.

Gooch had discovered that at 40 mph no less than a third of the power generated by a locomotive was required to move its own weight, while at 60 mph this proportion rose to a half. The atmospheric system seemed ideal as it necessitated no heavy engine. The novelty appealed to Brunel. Unfortunately his sense of perception deserted him – when the directors sought his opinion he was greatly in favour of adopting this system and reported:

> But it is in the subsequent working that I believe the advantages will be most sensible.
>
> In the first place with the gradients and curves of the South Devon Railway between Newton and Plymouth, a speed of 30 miles per hour would have been, for locomotives, a high speed, and under unfavourable circumstances of weather and of load it would probably have been found difficult and expensive to maintain even this; with the Atmospheric, and with the dimensions of pipes I have assumed, a speed of 40 to 50 miles per hour may certainly be depended upon, and I have no doubt that from 25 to 35 minutes may be saved in the journey.
>
> Secondly the cost of running a few additional trains, so far as the power is concerned, is so small, that plant of the engines, the attendance of the enginemen, etc., remaining the same, that it may almost be neglected in the calculations, so that short trains, or trains with more frequent departures, adapted in every respect to the varying demands of the public, can be worked at a very moderate cost.

Brunel proposed constructing an embankment over Cockwood Marsh between Starcross and Dawlish, floating the line on faggots, as Stephenson had done at Chat Moss on the Liverpool & Manchester Railway. The scheme failed to work here so a 200-yard-long timber viaduct was substituted, instead. The track used was his usual 62 pounds per yard bridge rail on longitudinal baulks.

Due to the cost of the pipe, he recommended single line; his rationale being that, because a head-on collision using atmospheric working was virtually impossible, there would be no danger. Brunel intended to use three different sizes of tube: 13 inches for Exeter to Newton Abbot; 22 inches for the steep gradients west of Newton Abbot and 15 inches for the rest of the line onwards to Plymouth. After 4,400 tons of 13-inch pipe had been delivered, Brunel decided that 15 inches would be better. The larger-diameter pipe required a more powerful pump, but inexplicably instead of increasing the size of the engines, he ordered auxiliary engines. Pumping stations were erected approximately every 3 miles along the line. A surprising feature – or perhaps oversight – was that, unlike the stations, the pump houses did not have telegraphic equipment, which meant that someone had to rush between station and engine house notifying any out-of-course delay.

The South Devon Railway between Exeter and Teignmouth was opened on 30 May 1846, using locomotive traction as the atmospheric system was incomplete. The first atmospheric trains ran on 13 September 1847 and were relatively fast and silent.

The highest speed recorded for an atmospheric train was 68 mph with a train of 28 tons. Brunel's son, later Dr Isambard Brunel, Chancellor of the Diocese of Ely, commented:

> The motion of the train, relieved of the impulsive action of the locomotive, was singularly smooth and agreeable; and the passengers were freed from the annoyance of coke-dust and the sulphurous smell from the engine chimney.

Restarting a train from an intermediate station was not quite so prompt as using a conventional locomotive. As an atmospheric train approached a station, the piston left the pipe by means of a self-acting valve. Dr Isambard Brunel continued:

> An arrangement for starting the train rapidly from the station, without help of horses or locomotives, had been brought practically into operation. This consisted of a short auxiliary vacuum tube containing a piston which would be connected with the train by means of a tow-rope and thus draw it along till the piston of the piston-carriage entered the main atmospheric tube. Some accidents at first occurred in using this apparatus, but its defects after a time were removed.

Elaborating on his father's work, he wrote:

> The Atmospheric System was vaguely credited with every delay which a train had experienced in any part of its journey; though, in point of fact, a large proportion of these delays were really chargeable to that part of the journey which was performed with locomotives. It often happened that time thus lost was made up on the Atmospheric part of the line, as is shown by a record of the working which is still extant. In the week 20–25 September 1847, it appears that the Atmospheric trains are chargeable with a delay of 28 minutes in all; while delays due to the late arrival of the locomotive trains, amounting in all to 62 minutes, were made up by the extra speed attainable on the Atmospheric part of the line.

In February 1848, Brunel updated the directors with another report:

> Notwithstanding numerous difficulties, I think we are in a fair way of shortly overcoming the mechanical defects, and bringing the whole apparatus into regular and efficient practical working, and as soon as we can obtain goods and efficient telegraphic communication between the Engine Houses and thus ensure proper regularity in the working of the Engines, we shall be enabled to test the economy of working. At present this is impossible, owing to the want of the telegraph compelling us to keep the Engines almost constantly at work, for which the boiler power is insufficient, and the consequence of this is that we are not only working the Engines at nearly double the time that is required, but the boilers being insufficient for such a supply of steam, the fires are obliged to be forced, and the consumption of fuel is irregular and excessive. There is every prospect of this evil being speedily removed, and as the working of the Atmospheric will then become the subject of actual experiment, and its value be practically tested, I shall refrain from offering at present any further observations upon it.

Atmospheric working continued for eight months until problems grew too great. The leather longitudinal seal at the top of the pipe between the rails was difficult to keep supple as the natural oil was sucked out by the pipe's inlet and outlet valves, and cold weather caused the leather to freeze. Originally a mixture of beeswax and tallow was rubbed into the leather, but this proved unsatisfactory in hot weather. It was replaced by a lime-soap-based composition, but this formed a hard skin, so was again replaced by cod oil and soap. This tended to be sucked into the pipe as it was insufficiently viscous and so needed constant replacing. Another problem was that the pumping engines frequently failed and the expenditure of fuel per indicated horsepower was more than double that of the best of the Cornish pumping engines to which they were analogous. To enable the engines to be worked more economically Brunel proposed installing an hydraulic accumulator at each pumping station. Train braking proved to be another problem as the handbrake on the piston carriage was sometimes overcome by the atmospheric pull on the piston.

In order to make level crossings on the atmospheric railway truly level, a special device was used. A hinged iron plate, connected to a piston in the cylinder, covered the atmospheric pipe and allowed road vehicles to cross the line. When the pipe was exhausted for the passage of a train the piston was sucked down, thereby raising the iron plate like a drawbridge. Following the passing of a train, the iron plate descended allowing clear passage for road vehicles. Unfortunately there was no device to prevent the contraption rising when a vehicle was crossing.

The start-to-stop time for a passenger train for the 20 miles from Exeter to Newton, with four intermediate stations, was fifty-five minutes. This poor average speed was due to having to wait for an oncoming train and then the additional delay of having to draw a train forward by horse or auxiliary piston until the main piston was again engaged in the tube.

The atmospheric railway was a complete disaster. The cost of installing atmospheric traction had been nine times Brunel's first estimate, while working costs were 3s 1d per mile compared with just 1s 4d by steam locomotive. When the system was abandoned, the South Devon Railway was left with 40 miles of single line which otherwise would have been made double, and steeper gradients than if designed for locomotive working. About the only benefit was its delightful engine houses with their Italianate features.

Daniel Gooch, the GWR locomotive superintendent and a fine engineer in his own right, was not convinced of the superiority of the atmospheric method, and wrote:

I went with Mr Brunel & several directors to Ireland to see the working of the system on the Dalkey line. From the calculations I then made of the power used to work the trains I found I could do the work much cheaper with locomotives. The result of our visit, however, was the determination to use the atmospheric pipes on the South Devon. I could not then understand how Mr Brunel could be so mislead [*sic*] as he was, and believe while he saw all the difficulties and expense of it on the Dalkey, he had so much faith in his being able to improve it that he shut his eyes to the consequences of failure, and there is no doubt the result of its trail on the South Devon between Exeter and Newton cost that company more money that would have well-stocked their whole line with locomotives and worked them also. The cost on this 30 miles of line was about £350,000. I may now say it was laid with all the care possible, and with all the improvements that could be suggested, and it miserably failed, and I was very soon called upon to supply locomotive engines to work the line.

I felt very sorry for Mr Brunel in this matter as it did him some harm, and figures showed it never had a chance to compete with the locomotive in cost and handyness [*sic*]. Mr Robt Stephenson made a very careful report on the working on the Dalkey line after we were there, and arrived at the same conclusion as I did. I may also mention a short piece was tried between London & Croydon which also was given up after a few months. This is certainly the greatest blunder that has been made in railways.

Daniel Gooch also reveals that Brunel was extremely worried about locomotives working on the South Devon Railway's gradients, which in places was as steep as 1 in 36:

> The South Devon line was opened in July this year 1847, between Newton & Totnes over long gradients of 1 in 42. I never saw Mr Brunel so anxious about any thing as he was about this opening. Relying upon the atmospheric principle he had made these steep inclines, and he feared there might be difficulties in working them. These difficulties disappeared with the day of opening. All our trains went through very well, and at night it seemed a great relief to Mr Brunel to find it was so. He shook hands with me and thanked me in a very kind manner for my share in the day's work. He never forgot those who helped him in a difficulty.

Atmospheric traction was abandoned on 10 September 1848. It says much for Brunel that after the atmospheric disaster and until the South Devon Railway was completed to Plymouth on 2 April 1849, he only accepted a nominal retaining fee. It is also to the credit of the directors that he continued to work for the company.

During the summer of 1847, while engaged on the South Devon Railway and consequently spending so much time in the area, his wife Mary insisted on a house being rented at Torquay for the summer and autumn. That September he bought land at Watcombe, midway between Torquay and Teignmouth overlooking Babbacombe Bay. From then on, every summer the family rented a house in the neighbourhood and Brunel always visited Watcombe whenever his business allowed. From that time to within a year of his death, the improvement of this property was Brunel's chief delight. He intended to build a house there, and drafted plans which never came to fruition. Had the house been built it would have had a marvellous view from the terrace with the sea on one side and Dartmoor on the other, while in front Torbay looked like a lake.

When he purchased the property it consisted of fields divided by hedges, but, assisted by William Nesfield, gardener at Eton, Brunel, always keen on detail, laid it out in plantations of trees and plotted the growth of every tree and bush. He designed tools for transplanting young trees and took care that a large amount of earth was moved along with a tree. He also designed a bridge of rough poles across the turnpike road between Teignmouth and Torquay. His favourite features were his Italian Garden, Mary's Clump, the Sea Walk with a double circle of twenty-one trees and 'mountain ashes everywhere I can stick them for the sake of their berries'. In 1854 he asked William Burn, a leading country house architect, to design a house in the French chateau style, but his large investment in the *Great Eastern* limited the funds available and the house was never built. At his own expense he erected a small church in Church Road.

Meanwhile, in 1846 a company had been formed for the construction of a dock at Millbay, Plymouth. A floating pier was erected in 1852 and the wet dock and graving dock in 1856. These features offered traffic to the railway.

In 1848, Brunel was able to purchase 17 Duke Street, a smaller adjoining house on the north side of No. 18 which he already possessed. Doorways were knocked between the houses making them into one expanded building. The lower floors of No. 18 were then occupied by the domestic staff. The 1851 census records seventeen people living there: Brunel and Mary, their children Isambard, Henry and Florence, his widowed mother Lady Brunel, a governess and two nurses, butler, housekeeper, cook, lady's maid, three housemaids and a kitchen maid. He rebuilt the property so that he was able to extend his offices on the ground floor, while above he created a large dining room, known as the 'Shakespeare Room' as on the walls were displayed scenes from the writer's plays which Brunel had commissioned from eminent painters; his collection included Landseer's *Titania*. The large dining table contained huge silver-gilt centre and side pieces presented by the Great Western Railway Company. This room was considered the last word in sumptuous taste. One drawing room contained a chamber organ. His houses backed on to St James's Park, but Duke Street has since disappeared below government offices. It ran parallel with Horse Guards Street, from Great George Street almost as far as Downing Street.

His granddaughter, Lady Celia Noble, tells us more about Duke Street in *The Brunels, Father and Son*:

The rooms lent themselves readily to the private theatricals so much loved by the family, and permitted more ambitious efforts than at High Row. There were 'tableaux' of scenes from Flaxman's *Odyssey*, in which Mary, with her classical features and stately figure, represented Penelope with her suitors, using an elaborate loom made by Webbe, the famous upholsterer of Bond Street; and there were frequent cathedral interiors with appropriate music from Mendelssohn on the organ in the back drawing-room; Isambard, now the important business man, no longer played farcical parts, but in spare moments designed mechanical contrivances for the scenery.

II

ANOTHER TUNNEL, AN EXHIBITION AND PADDINGTON STATION
1847–52

In April 1848 the panic caused by the second French Revolution and the possibility of disorder in Britain caused Brunel to be enrolled as a special constable for Westminster. Brunel acted as one of the two 'leaders' of the district between Great George Street and Downing Street. It was also in April 1848 that Brunel visited Paris with his brother-in-law John Horsley.

A recession of the same year caused Brunel to have to downsize the number of his assistant engineers. He found this painful and in August wrote:

I have generally anxieties and vexations of my own, and at present they are certainly not *below* the average but they are completely absorbed and overpowered by the pain I have had to undergo for others. For some weeks past, my spirits are completely floored by a sense of the amount of disappointment, annoyance, and – in too many cases – *deep distress* inflicted by me, though I am but an instrument, in dismissing young men who have been looking forward to a prosperous career in their profession, unsuspicious of the coming storm which I, and others mixing in the world, have foreseen. It is positively shocking to see how many of these young engineers have looked upon their positions as a certainty, have been marrying and making themselves happy, and now suddenly find themselves in debt and penniless. You can hardly imagine what I have to undergo in receiving letters of entreaty which I have no power to attend to. Everywhere we are reducing.

In October 1849, the GWR's Slough to Windsor branch opened, crossing the Thames by a 203-foot-span wrought-iron arch-and-girder bridge, the outward thrust of the arch ribs contained by the wrought-iron deck girders, and abolishing the need for the abutments to support a horizontal load. The abutments supporting the girders were six 6-foot-diameter cast-iron cylinders. The bottom sections were given a cutting edge, and the cylinders were sunk to a solid base by excavating their interior and placing weights on top to force them into the ground – a similar procedure to that used by the senior Marc when sinking the Rotherhithe shaft of the Thames Tunnel. Brunel used a similar design for the bridge at Newport which replaced the timber viaduct destroyed by fire on 31 May 1848.

At Llansamlet between Neath and Swansea a serious landslide blocked a deep cutting. To resolve the problem without resorting to easing the cutting slopes, Brunel constructed four flying arches weighted with locally obtained heavy copper slag to counter the inward thrust of the cutting sides.

Brunel's father suffered a stroke on 7 November 1842 and died, aged eighty, on 12 December 1849. He was buried in Kensal Green Cemetery. His long life in the service of public works had been justly recognised by the knighthood from Queen Victoria. His wife Sophia moved in with her son Brunel, but died eventually in January 1855 and was buried with her husband.

Brunel's letters of the period show that his business matters did not cease in needing to be addressed. His strong character, insisting that he would only accept a post where he alone was responsible, was still very much in evidence and apparently unperturbed in a letter of 30 December 1851:

I shall be happy to act in any capacity (subject to the exception I will further explain) which can be useful to your company; but the exception I have to make is one which perhaps resolves itself merely into a question of *name*. The term 'Consulting Engineer' is a very vague one, and in practice has been too much used to mean a man who for a consideration sells his name, but nothing more. Now I never connect myself with an engineering work except as the Directing Engineer, who, under the Directors, has the sole responsibility and control of the engineering, and is therefore 'The Engineer' and I have always objected to the term 'Consulting Engineer'.

The 887-yard-long Mickleton Tunnel, later known as Campden Tunnel, on the Oxford, Worcester & Wolverhampton Railway involved a labour dispute. Although started in 1846, in June 1851 the contractor had a dispute with the company and unsurprisingly ceased work as he had received no payment for his outstanding efforts. The OWWR took possession and handed the contract to Messrs Peto & Betts. The original contractor, Robert Marchant, Brunel's second cousin, objected and kept his men on guard. Brunel then arrived with an army of navvies to take possession of the site. Marchant, expecting a fight, asked two magistrates to attend:

> The magistrates were early on the ground, attended by a large body of police armed with cutlasses. Mr Brunel was there with his men, and Mr Marchant, the Contractor, also appeared at the head of a formidable body of navigators. A conflict was expected, but happily through the prompt action of the magistrates, who twice read the Riot Act to the men, they were dispersed.

On 17 July 1851, Brunel, Robert Varden, his resident engineer at Banbury, and Mr Hobler, Brunel's legal adviser, set out in the 'Flying Hearse' at the head of a 2,000-strong crowd of Peto & Bett's men. They had been assembled from various sections of the line, and marched that night to the tunnel, arriving in the early hours of the following morning. Brunel was rebuked by the magistrates for this provocation, but was not arrested for inciting a riot. While Hobler was talking with them, a fight erupted among the navvies. One man produced a brace of pistols, but was knocked to the ground. One magistrate read the Riot Act twice before Brunel called off his navvies. Brunel returned just after dawn with a small group of men, hoping to surprise the opposition, but they were still on guard. Fighting again broke out and the Riot Act was again read. Late on the following evening of 18 July 1851, Brunel reassembled his 'army' of 2,000 men. A great fight ensued; a magistrate pointed out to Brunel that he was responsible for any injuries and he agreed to keep the peace. Marchant and his hundred navvies, overawed by the 2,000-strong Brunel faction, surrendered on a promise of an arbitrated settlement. The Battle of Mickleton Tunnel was over. Troops from Coventry, who had been sent for to back up the police, arrived to find all was quiet.

Marchant was declared bankrupt on 29 November 1851. He wrote Brunel an abusive letter to which he received the reply on 16 February 1852:

> Marchant, I can't make out why, with such feelings as you profess to entertain, you address me at all. You go into partnership with a man who has no capital but borrowed money and who is involved in an unfortunate contract and you are ruined. How could it be otherwise? And then you abuse the only man who ever did or does care sixpence for your welfare. You now talk of *forgiving* this man. This is cant. Your excessively irritable temper, your impudence, I can stand, but not your cant. As long as you wish to abuse me, have the goodness to do it in some other way than writing to me as I shall return the letters unread.

Robert Marchant's father, William Marchant, kept a farm at Chilcompton, south of Bath. Marc Brunel and his wife Sophie had holidayed there in 1841 and 1843 to give him a well-earned break from his work. It was when they were staying there in 1843 that Marc, to his great disappointment, missed the surprise visit of Queen Victoria and Prince Albert to the Thames Tunnel.

As the OWWR increasingly became antagonistic towards the GWR, in March 1852 Brunel felt he had to resign from being the

OWWR's engineer, instead being replaced by John Fowler, who later designed the Forth Bridge.

Although some broad gauge rails were laid on the OWWR in order to comply with the Act's legal stipulation that it be a broad gauge line, they were never used for regular traffic as the company managed to evade its undertaking to run broad gauge trains. The OWWR then joined forces with two other standard gauge lines, the Worcester & Hereford Railway and the Hereford, Aberavon & Newport Railway, to become the West Midland Railway. When in 1863 the West Midland was taken over by the GWR and it became essential to work through traffic over this line to and from London, the line between Oxford and Paddington had to become mixed gauge. Thus, four years after Brunel's death, standard gauge finally reached Paddington. For the broad gauge it was the beginning of the end of this lengthy contest.

Work began in 1851 on a 13-acre dock complex designed by Brunel for Millbay, Plymouth. The entrance gates were of wrought iron and weighed over 75 tons. They were designed with large air chambers to make them buoyant. In 1852 he designed a 300-foot-long pontoon for coaling ships. It had a capacity of 4,000 tons of coal and was connected to land by a twin-span bridge.

The crisis brought about by the 1846 Irish famine and its subsequent effect on the economy meant that in 1851 work on the South Wales Railway was halted before reaching Fishguard. Brunel sensibly minimised his losses by abandoning the plan for a line from Haverfordwest to Fishguard and changed the destination port to Neyland; the extension from Haverfordwest was opened on 15 April 1856.

In Brunel's report for August 1856 to the South Wales Board he wrote:

The success of the Steam Boat communication with Ireland has shown that a portion, if not the whole, of the Floating Pier, as originally designed, is not only desirable but essential for carrying on the trade, particularly in Cattle, which promises to be

considerable. Accordingly a portion has been ordered and will be
immediately proceeded with.

The pontoon he referred to was a wrought-iron affair brought into
use in the spring of 1857. Two years later it was extended by the
addition of four more pontoons which Brunel had previously used
for floating the piers of the Royal Albert Bridge.

Brunel was involved with the Great Exhibition of 1851 as a
member of the Machinery Section Committee and the Building
Committee. He was also chairman and reporter of the judges for
Class 7 on civil engineering, architecture and building contrivances.
On 11 March 1850, he wrote a letter strongly opposing giving
prizes in the Machinery Section. His views were not shared by his
fellow members, but his inclination was evidently shared when no
prizes were awarded at the Paris Universal Exhibition in 1867.
It was probably due to his influence that Gooch's broad gauge
Lord of the Isles was displayed at the Great Exhibition on a plinth
high above the standard gauge engines.

None of the 245 designs for a hall for the 1851 exhibition were
accepted by the Building Committee, which decided to produce
its own. This consisted of low brick buildings surmounted by
an iron dome 200 feet in diameter and 150 feet high – the dome
being Brunel's contribution. Eventually, Joseph Paxton produced
his inspired iron and glass Crystal Palace; it was Paxton's work
which influenced Brunel in turn when he designed the new
passenger station at Paddington. When the Exhibition ended
Brunel was compelled, much against his will, to accept a pecuniary
acknowledgement for his services. He spent the money erecting
model cottages at Barton, a village near his property in Devon.

Paxton planned that at the end of the Great Exhibition the
Crystal Palace should be retained in Hyde Park as a winter park
and garden under glass. When this scheme was rejected, he formed
a company to purchase the building and re-erect it on another site.
The site chosen was Sydenham Hill, Croydon, where it reopened
on 10 June 1854.

A large water tank was required to heat the building and supply the fountains. As chimneys were required at each end of the structure, it was thought that a chimney could double as a water tower. These were designed by Paxton's assistant, Charles H. Wild. Paxton felt slight misgivings about the proposed structures and sought Brunel's advice. He said that 'the attempt to support upwards of 500 tons at a height of more than 200 feet upon a cluster of slender legs with but a small base involves considerable difficulties'. He further observed that the twelve main legs should not double as water pipes and that the legs should be braced to prevent buckling; and he opined that the water tank be of wrought rather than cast iron, and its weight carried by the columns rather than through supporting struts. The concrete foundations were unsatisfactory and needed replacement.

Brunel superseded Wild as engineer responsible for the towers. Then, when the towers were actually under construction, Paxton decided to increase that tank's capacity to 1,500 tons. As the towers were laid on a 1 in 12 slope, each foundation was a concrete ring of 58-foot diameter sunk 10 feet deep. This ring supported brickwork 18 feet high, on which were fixed cast-iron foundation plates to support cast-iron columns to carry the base of the water tank. A brick chimney rose through the centre of each tower. The opening ceremony was on 18 June 1856.

Although most of Brunel's bridges were beautifully symmetrical, the structure over the Wye at Chepstow was a notable exception due to the east side having a 120-foot-high cliff and the west bank low-lying land. A single span of 300 feet crossed the river giving 50 feet of clearance at high tide. The cliff provided an abutment for one end of the span, but the other end, together with the three approach spans were carried on piers.

Brunel used cast-iron cylinders of 8-foot and 6-foot diameter respectively, which were sunk through the soft ground to a solid foundation and then filled with concrete. Brunel had an experimental cylinder made with an external screw thread so that it could be driven home like a screw pile. Although it worked well in gravel and clay, the beds of running sand did not allow it to grip and the idea was abandoned. Water in these sand beds caused trouble. To sink the large columns for the main span Brunel used the pneumatic method, whereby the upper ends of the cylinders were sealed and compressed air introduced to keep the water out, workmen and materials of necessity then entering through airlock doors.

The river span consisted of girders supported by suspension chains hung from a horizontal circular tube of 9 feet diameter. These tubes, very slightly arched to improve their appearance, prevented the chains dragging the piers inwards. He had first used a truss of this type in 1849 for a road bridge spanning the entrance to Cumberland Basin in Bristol. That Brunel took the utmost care to avoid accidents is shown by the fact that one span was assembled on temporary piers on the river bank and tested beforehand with a distributed load of 770 tons.

On 8 April 1852, a tube was floated into position below the piers and then lifted by chain tackle attached to specially designed winches. Brunel was in charge, assisted by R. P. Brereton and Captain Claxton, the latter in charge of the vessel side of the operation.

The first tube was placed at right angles to the Wye and thrust forward on trolleys until it overhung the river; beneath was a pontoon of six iron barges. The pontoon took the weight of the span as the tide rose and was guided across the river by two chains anchored to the opposite cliff. By sunset the tube had been lifted to rail level and the next day was set on its piers. A single-line bridge opened on 19 July 1852 and the second followed on 18 April 1853. This bridge at Chepstow proved to be something of a dummy run for the Royal Albert Bridge at Saltash.

Brunel also used a lattice girder design in his railway bridge at Windsor and his road bridge over the River Dee at the entrance to Balmoral Castle; both of these bridges are still extant today. The Balmoral bridge is a single-span, wrought-iron, plate-girder construction, slightly cambered. The two riveted, wrought-iron girders supported on masonry piers give a span of 125 feet. He incorporated umbrella-shaped flanges into the girders top and bottom so that the rain would run off the iron rather than corrode it. The original timber deck was replaced in 1971 and more recently received Permadeck timber panels. The maker's plates read: 'R. BROTHERHOOD CHIPPENHAM WILTS 1856.' Although the metal work had been completed by Brotherhood in 1856 and despatched to Scotland, due to the work on the piers being tardy, erection was not completed until the autumn of 1857. Unfortunately the royal family was not delighted with its functional, rather than Gothic, appearance. Receiving this criticism, Brunel replied to Colonel Charles Phipps, Keeper of the Privy Purse:

> I am much disappointed at your report of the appearance of the bridge at Balmoral. I confess I had hoped for a very different result and thought that at all events the perfect simplicity of the construction and absence of any attempt at ornament would secure it from being in any way unsightly or offensive, which I think is always a great first step, but I fear your expression of not extremely ornamental implies something very much the reverse.
>
> As regards the elasticity I trust that it is not felt to an extent that is unpleasant as it is unavoidable ... But with regard both to the appearance and the stiffness, I will take an early opportunity of seeing it myself.

In Australia, the State of Victoria, newly created in 1850, sought help from Brunel. Saltwater Creek, west of Melbourne, was crossed by a wrought-iron bridge based on the one at Balmoral. On 29 May 2006, HRH Prince Philip, Duke of Edinburgh, unveiled a plaque on the original acknowledging Brunel as the designer.

Back in London, the original Paddington station had been but a plain and temporary terminus due to a shortage of capital. It originally used the arches of the nearby Bishop's Bridge, and Brunel had always yearned for something much grander as an entrance to his Great Western Railway. As the new Paddington station was based on the Crystal Palace design, Brunel believed it should be constructed by Fox, Henderson & Company, who had been responsible for the building which was to hold the Great Exhibition. On 2 January 1851, Brunel produced outline plans to the Board. Shortly thereafter, on 13 January 1851 he wrote excitedly to the architect Matthew Digby Wyatt:

> I am going to design, in a great hurry, and I believe to build, a Station after my own fancy; that is, with engineering roofs, etc. etc. It is at Paddington, in a cutting, and admitting of no exterior and all roofed in. Now, such a thing will be entirely *metal* as to all the general forms, arrangements and design; it almost of necessity becomes an Engineering Work, but, to be honest, even if it were not, it is a branch of architecture of which I am fond, and, *of course,* believe myself fully competent for, but for *detail* of ornamentation I neither have the time nor knowledge, and with all my confidence in my own ability I have never any objection to advice and assistance even in the department which I keep to myself, namely the general design.
>
> Now, in this building which, *entre nous,* will be one of the largest of its class, I want to carry out, strictly and fully, all those correct notions of the use of metal which I believe you and I share (except that I should carry them still farther than you) and I think it will be a nice opportunity.
>
> Are you willing to enter upon the work *professionally* in the subordinate capacity (I put it in the least attractive form first) of my *Assistant* for the ornamental details? Having put the question in the least elegant form, I would add that I should wish it very much, that I trust your knowledge of me would lead you to expect anything but a disagreeable mode of consulting you, and of using and acknowledging your assistance; and I would remind you that it may be as good an opportunity as you are likely to have (unless it leads to others, and I hope better) of applying that principle you have lately advocated.
>
> If you are disposed to accept my offer, can you be with me this evening at 9 ½ p.m.? It is the only time this week I can

appoint, and the matter presses *very much*, the building must be half finished by the summer. Do not let your work for the Exhibition prevent you. You are an industrious man, and night work will suit me best.

I want to show the public also that *colours* ought to be used. I shall expect you at 9½ this evening.

Brunel provided the station's broad outline, Digby Wyatt attending to the details. Brunel provided the splendid roof – the first British station to have a metal roof. *The Life of Isambard Kingdom Brunel,* by his son, describes it well:

The interior of the principal part of the station is 700 ft long and 238 ft wide, divided in its width by two rows of columns into three spans of 69 ft 6 in, 102 ft 6 in and 68 ft, and crossed at two points by transepts 50 ft wide, which give space for large traversing frames. 'the roof is very light, consisting of wrought iron arched ribs, covered partly with corrugated iron and partly with the Paxton glass roofing, which Mr Brunel here adopted to a considerable extent. The columns which carry the roof are very strongly bolted down to large masses of concrete, to enable them to resist sideways pressure.

This station may be considered to hold its own in comparison with the gigantic structures which have since been built, as well as with older stations. The appearance of size it presents is due far more to the proportions of the design that to actual largeness of dimension. The spans of the roof give a very convenient subdivision for a large terminal station, dispensing with numerous supporting columns and at the same time avoiding heavy and expensive trusses. The graceful forms of the Paddington station – the absence of incongruous ornament and useless buildings – may be appealed to as a striking instance of Mr Brunel's taste in architecture and of his practice of combining beauty of design with economy of construction.

The *Illustrated London News* for 5 August 1854 sketches further details:

> The roofing contains 189 wrought iron ribs, or arches, of an elliptical form, and arranged in rows of three each, parallel to one another, with twelve diagonal ribs at the transepts. The height to the under side of the ribs in the central opening is 54 ft 7 in from the line of rails: from the springing it is 33 ft 9 in. The height of the side divisions is 46 ft the ribs against the building are supported by sixty-three cast-iron pilasters, or square columns, and those to the other portion of the roofing rest on sixty longitudinal wrought-iron girders, each about 30 ft long, and supported by sixty-nine circular columns. The ribs over the central opening are 1 ft 8 in high, formed of quarter-inch plate with flanges top and bottom, giving a width there of 6 inches. The ribs over the side openings are 1 ft 4 in high, formed in the same manner. The central half of each of the curved roofs is glazed, and the other portion is covered with corrugated galvanised iron.
>
> There are two transepts – one a third and the other two-thirds down the station. Facing the length of each transept is a balcony on the office block.
>
> The execution of the design has been superintended by Mr Brunel, principally through one of his chief assistants, Mr Charles Gainsford. The work was done by Messrs Fox, Henderson and Co.

The principal buildings, 580 feet in length and 30–40 feet in width, were alongside the departure platform; management departments used the upper floors and the traffic department the lower ones. As Queen Victoria used the station frequently when en route to Windsor, a suite of waiting rooms was provided especially for her use, with the convenience of direct access from the carriage approach to the station platform.

The station cost £650,000 and covered 8 acres. Its interior was originally painted red and grey. Unlike the other major London termini, Paddington has no imposing exterior façade as the train sheds are fitted into a cutting. The fact that the station was designed by Brunel rather than an outside architect meant that the GWR saved some £20,000 to £30,000 in fees.

It was not only large engineering projects which interested Brunel. In October 1852 he wrote to Westley Richards, a gunsmith: 'I have long wanted to try an experiment with a rifle, for the purpose of determining whether there is anything in a crochet I have upon the subject.' His notion was that a gun would prove more accurate if a shot was given spin. He intended achieving this by using a barrel of an octagonal cross-section with the twist increasing from breech to mouth.

Although this was an important development in gunnery, Brunel did not take out a patent. This did not prevent Joseph Whitworth from doing so after he had seen the rifle in Richards' shop. In 1858, Richards observed that Brunel's rifle infringed Whitworth's patent. Brunel said: 'I have never seen Whitworth's patent. What is it exactly he does patent? It cannot be the polygon.' Brunel said he had no intention of challenging 'my friend Mr Whitworth'. From that small incidental beginning, with a design of no great import to Brunel, grew Whitworth's great armament business.

From 1 May 1849, the Bristol & Exeter Railway ceased to be worked by the GWR and ran its own affairs. This placed Brunel in a conflict of interests, so he resigned as engineer to that company. Until 1854 he held the post of consulting engineer; immediately following his resignation, one of his pupils, Francis Fox, was appointed engineer and held that position until the company was taken over by the GWR in 1876.

Brunel travelled to Venice in 1852 and bought pictures to add to his growing collection. In fatherly mode, and indeed now as head of the family, he sent his fifteen-year-old son Isambard the following letter on 22 November 1852:

How for arithmetic? I think I must take you in hand at Christmas for I fear very much that the quality of your arithmetic knowledge is very queer, whatever the quantity and that is probably small.

Half a pint of poor small beer is not nourishing and cannot be called a malt liquor diet, neither can I imagine Harrow arithmetic to be much better. It is very distressing but one must put up with it as one would if brought up in a country where it was the practise to put out one's eye.

12

A ROYAL BRIDGE, A SUBMARINE AND A HOSPITAL 1852–56

In 1852, regarding the bridge across the Tamar at Saltash, Brunel reported to the Board of Trade:

> This bridge has been always assumed to be constructed for a double line of railway as well as the rest of the line. In constructing the whole of the line at present with a single line of rails, except at certain places, the prospect of doubling it hereafter is not wholly abandoned, but with respect to the bridge it is otherwise.
>
> It is now universally admitted that when a sufficient object is to be attained arrangements may easily be made by which a short single line can be worked without any appreciable inconvenience ... this will make a reduction of at least £100,000.

The contract for building the Royal Albert Bridge was let in January 1853 and the foundation of the first of the land piers laid on 4 July 1853. The resident engineer was Brunel's chief assistant, Robert Pearson Brereton. For constructing the central pier, Brunel decided to build it inside a wrought-iron cylinder similar to that used for the exploratory work, but very much upscaled to form a combination of a diving bell and a compressed-air caisson. It weighed 300 tons and was floated into position and sunk in

June 1854. The main part of the cylinder measured 37 feet in diameter and was 90 feet high. The lower part of this cylinder had a diving bell compartment 4 feet in width which could be pressurised.

A second cylinder, 10 feet in diameter, extended from the diving bell roof to provide air and access. Within this was the 1848 original, 6-foot-diameter cylinder to give access through airlocks to the outer compartment kept under air pressure. This dual arrangement was designed to avoid workmen engaged on building the piers the disadvantage of working under conditions of dangerous air pressure.

Difficulties were encountered in sinking the large cylinder through dense beds of oyster shells, and rock was not reached until February 1855. Then another problem occurred. A fissure in the rock let water into the diving bell, where it overcame the onboard pumps; the air pressure in the outer compartment had the effect of forcing the water through this fissure into the diving bell.

Brunel's solution was to attempt to re-pressurise the diving bell, but the 10-foot-diameter cylinder had not been constructed to withstand such pressure, so yet another cylinder, 9 feet in diameter, was installed within the 10-foot cylinder. This was the first use of a large compressed-air caisson.

Brunel's unfortunate experience with the Italian railway being reluctant to pay him did not put him off other foreign projects. In 1853 he prepared for the New Orleans & Great Western Railway a *Memorandum of Data Required in Reference to the Crossing of the Berwick Creek* which offered technical advice on pile-driving.

Brunel's continuing concerns for safety is shown by the letter of 29 November 1853 to Charles Saunders, the GWR secretary:

The accidents by truck flaps seems to be on the increase although the numbers of defective pins are certainly on the decrease – an accident happened at Melksham on Saturday [26 November] doing damage and endangering seriously the goods shed which can hardly have arisen from defective points unless they are positively alive – for the train had not left the station when a flap which had just been well keyed up fell down and carried away part of one end of the goods shed.

With respect to the point being the cause – had it struck you a singular circumstance that since the opening of the railway we have never heard a flap dropping on the off side – the side not next to the small station platform. It would be a most serious thing if it did happen as it would cut up a horse box or the Queen's Carriage on the other line. In fact the consequences might be so fearful that I have kept it to myself lest the hint might lead to the result – but it shakes my belief in the effect being caused by the pins.

At the time of the Crimean War, around 1854–5, Brunel took up the question of improving large guns, a subject which was attracting the attention of several scientists interested in contributing to the national cause. His friendship with William Armstrong of Newcastle-on-Tyne gave him the opportunity of discussion and the chance of his ideas being carried into effect. He made several innovative suggestions to the British government. One scheme consisted of making a gun barrel with wire wrapped around it to prevent it bursting, but they were obliged to abandon the project when the idea was patented in May 1855 by Mr Longridge.

In September 1854, Brunel designed a semi-submersible gun battery which would offer only a minimum target to the enemy. The ammunition was kept out of harm's way in the submerged hull, with the 12-inch gun firing three rounds a minute via a breech-loaded mechanism. To avoid the fouling of a screw or rudder, the hull was propelled by three steam jets, one for short-distance propulsion and two for lateral manoeuvres – these jets being under the control of the gun-layer. Although screw propulsion would have been possible, he preferred a jet as this solution exposed nothing which could be damaged by an enemy shot. For longer-distance transportation, Brunel wrote: 'I propose a ship of the class of small screw colliers made to open at the bows and its contents floated out ready for action.'

Brunel, remembering his unfortunate experiences with the Admiralty, refused to suggest his ideas to them, but a friend, Captain Christopher Claxton, removed plans and models from Duke Street and took them to the Admiralty. Brunel was not surprised that his intriguing idea was not adopted. He expressed to General Sir John Burgoyne his opinion of the Admiralty, namely that it was old-fashioned and refused to accept new ideas:

> You assume that something has been done or is doing in the matter which I spoke to you about last month – did you not know that it had been brought within the withering influence of the Admiralty and that (of course) therefore, the curtain had dropped upon it and nothing had resulted? It would exercise the intellects of our acutest philosophers to investigate and discover what is the powerful agent which acts upon all matters brought within the range of the mere atmosphere of that department. They have an extraordinary supply of cold water and capacious and heavy extinguishers, but I was prepared for and proof against such coarse offensive measures. But they have an unlimited supply of some *negative* principle which seems to absorb and eliminate everything that approaches them … It is a curious and puzzling phenomenon, but in my experience it has always attended every contact with the Admiralty.

At this period Brunel's time was also occupied with the *Great Eastern* and the Royal Albert Bridge and he made no further developments in his military designs. Later, after Brunel's death, Claxton went to the Admiralty to collect the model. It could not be found. At last, recollection came to one official, 'Oh, *I* know,' he exclaimed, 'it is a duck-shooting thing, is it not, painted white?'

In November 1854, aged forty-eight, Brunel observed that he had been responsible for constructing 1,046 miles of railway. Since the Great Western Railway contract had been let just over eighteen years before, he had completed an average of 58 miles of track annually – a great achievement considering his very close input and involvement with all of the many details.

Brunel's mother, Sophie Brunel, died on 5 January 1855.

In an interesting turn of events, Florence Nightingale's arch-villain was Benjamin Hawes, Brunel's brother-in-law, who had married his elder sister, Sophia, and was his nearest and dearest friend. Hawes, Permanent Under Secretary at the War Office, wrote to Brunel on 16 February 1855 to ask if he would be willing to design a prefabricated hospital for the Crimea which could be built in England and then shipped out. Brunel replied the same day: 'This is a matter in which I think I ought to be useful and therefore I need hardly say that my time and my best exertions without any limitations are entirely at the Service of the Government.' Only six days later he was asking for contoured sketch maps of the suggested sites, and an initial contract for the building of a 1,000-bed hospital had already been placed.

The War Office Contracts Department was much displeased at this rapid initiative and Brunel retorted:

Such a course may possibly be unusual in the execution of Government work, but it involves only an amount of responsibility which men in my profession are accustomed to take … It is only by the prompt and independent actions of a single individual entrusted with such powers that expedition can be secured and vexatious and mischievous delays avoided … These buildings, *if wanted at all*, must be wanted before they can possibly arrive.

Within a few days an experimental ward had been erected on GWR premises at Paddington. Each unit comprised two wards each for twenty-four patients and was completely self-contained with its own nurses' rooms, water closets, and outhouses, 'so that by no accident can any building arrive at its destination to be erected without having these essentials complete'. Each patient had 1,000 cubic feet of air and one large ventilator fan was provided for each unit. This forced air into the wards so that no smells could be drawn from the adjoining closets into the wards. Wash basins and invalid baths had wooden trunk drains. Surgery, dispensary and officers' quarters were in the same standard wooden units, but the kitchen, bake-house and laundry were in metal buildings to avoid any fire risk. Candles provided heating and lighting. The timber roof was covered with tin to reflect the heat of the sun. For the occasions when no breezes came through the windows, wards were ventilated by a man-operated fan. Water closets were provided – a luxury few patients would have enjoyed at home – while the wooden sewers were 15-inch square and tarred inside. Building parts were shipped to Turkey and erected at Renkioi on the southern shores of the Dardanelles.

All parts of the prefabricated hospital were arranged for transport in separate packages requiring two men, and each ship taking out the material carried one set of buildings, complete with pumps, mains and tools so that should one ship be wrecked accommodation would still be available for immediate use. The cost of the buildings delivered ready for shipment was around £18 to £22 per bed.

Later that March, Hawes was warned that 1,800 tons of shipping space would be shortly required in April. Brunton, engaged on building the Dorchester to Weymouth line, received a telegram one evening at his home in Dorchester summoning his presence at Brunel's Duke Street office at 6.00 a.m. the next morning. Brunton just had time to catch the night mail. Brunton himself recorded:

A footman in livery opened the door, and told me in reply to my enquiry that Mr Brunel was in his office room expecting me. I was ushered into the room blazing with light, and saw Mr Brunel sitting writing at his desk. He never raised his eyes from the paper at my entrance. I knew his peculiarities, so walked up to his desk and said shortly 'Mr Brunel I received your telegram and here I am'. 'Ah,' was his reply, 'here's a letter to Mr Hawes at the War Office in Pall Mall, be there with it at 10 o'clock.' He resumed his writing and without a further word I left his office.

This episode shows something of Brunel's work ethic and demeanour in professional affairs – it was even unusual for a professional man to be at work at such an early hour.

On 2 April 1855, Brunel, leaving nothing to chance, wrote to Brunton:

> All plans will be sent in duplicate ... By steamer *Hawk* or *Gertrude* I shall send a derrick and most of the tools, and as each vessel sails you shall hear by post what is in her. You are most fortunate in having exactly the man in Dr Parkes that I should have selected – an enthusiastic, clever, agreeable man, devoted to the object, understanding the plans and works and quite disposed to attach as much importance to the perfection of the building and all those parts I deem most important as to mere doctoring.
>
> The son of the contractor goes with the head foreman, ten carpenters, the foreman of the W.C. makers and two men who worked on the iron houses and can lay pipes. I am sending a small forge and two carpenter's benches, but you will need assistant carpenters and labourers, fifty to sixty in all ... I shall have sent you excellent assistants – try and succeed. Do *not let anything induce you to alter the general system and arrangements that I have laid down.*

Brunel again wrote to Brunton on 13 April 1855:

> Materials and men for the whole will leave next week. I will send you bills of lading for the five vessels: the schooner *Susan* and the barque *Portwallis*, the sailers *Vassiter* and *Tedjorat* and the *Gertrude* and *Hawk* steamers. By the first named steamer, a fast one, the men will go with Mr Eassie's son.

Concerned about cleanliness and aware that most of the soldiers would never have seen a water closet, he added:

> I would only add to my instructions attention to closet floors by paving or other means so that water cannot lodge in it but it can be kept perfectly clean. If I have a monomania it is a belief in the efficiency of sweet air for invalids and the only point of my hospital I feel anxious about this.

On 18 April 1855, he forwarded the bills of lading for the complete hospital and added:

> I trust these men will pull altogether, but good management will always ensure this – and you must try while you make each man more immediately responsible for his own work to help each other – and to do this it is a good thing occasionally to put your hand to a tool yourself and blow the bellows or any other inferior work, not as a display but on some occasion when it is wanted and thus set an example. I have always found it the answer.

In the meanwhile Brunel had written to Dr Parkes, the chief medical superintendent at Renkioi:

> All the vessels with the entire hospital will I believe have left England before the end of next week, that is before the 21st. Finding that none of the Ordnance Stores were likely to be ready, and indeed that no positive time could be ascertained for their being ready, I obtained authority yesterday to purchase one third of the required quantity of bedding and some other similar stores and they are now going aboard with the buildings. I have added twenty shower baths, one for each ward and six vapour baths. You will also be amazed to find also certain boxes of paper for the water closets – I find that at a cost of a few shillings per day an ample supply could be furnished and the mechanical success of the W.C.'s will be much influenced by this. I hope you will

succeed in getting it used and not abused. In order to assist in this important object I send out some printed notices or handbills to be stuck up, if you see no objection, in the closet room opposite each closet exhorting the men to use the apparatus properly and telling them how to do so. If you do not approve of such appeals the paper can be used for other purposes and perhaps impart some information in its exit from this upper world.

The buildings will be very quick after you; I almost fear you cannot have satisfied yourself about the site by the time they arrive. If you depend on Government Officers and if they at all resemble those at home, with one or two exceptions, your patience will be well tried.

The hospital's erection started on 21 May 1855. It took longer than expected because Brunton was unable to find any trustworthy local labour and was forced to use just the gang of eighteen sent from England. A railway ran from the jetty to the hospital and, had the war continued, a branch would have run along the corridor to the entrance of each ward. The covered corridor was left open in summer to allow a cool walk while in winter it was boarded up on the north side. By 12 July 1855 the hospital admitted 300 patients and by 4 December 1855 had its full quota of 1,000. No more patients arrived after 11 February 1856.

In the short time the hospital was in use before peace was declared it treated about 1,331 men, of whom only fifty died – a four per cent fatality rate compared to forty-two per cent at Scutari, though as Renkioi was further from the Crimea most of its patients were convalescent rather than recently wounded. At Scutari, for every one who died of wounds, three died from disease.

When peace came, Brunel sent Brunton instructions for the hospital's disposal:

Don't want the thing to be flung into a ditch when done with, but should prefer a useful end; that each part should be made the *most* of and methodically and profitably disposed of. Everybody here expresses themselves highly satisfied with everybody there and what we have done. I should wish to show that it was no *spirit* but just a sober exercise of common sense.

On young Isambard's eighteenth birthday, 20 may 1855, Brunel sent a kind and fatherly letter:

My dear Isambard,
You are now eighteen, getting into manhood, and I, of course, going down the other side of the hill.

May you have as much happiness as I, your father, have had; and try and remember always that at least half the evils of life – and those by far the most difficult to bear – are of one's own creation, either by follies and imprudences, or by obstinate and wilful omission.

And, my dear Isambard, although my constant engagements have prevented my seeing so much of my children as I should have wished, yet I hope that you would look upon me as your first friend to consult, if ever you got into difficulties or had any doubt as to your proper course.

Pray then, dear boy, while you have a Father, to whom you might safely and without annoyance to yourself, confide anything, consult him if you ever have the slightest difficulty; and to you, my dear Isambard, many happy returns of the day.
    Your affectionate father,
    I.K. Brunel

From 1855, Brunel was consultant engineer to the Eastern Bengal Railway in India. Towards the end of 1858, when his health forced him to take a holiday in Egypt, he considered the possibility of continuing to Calcutta but thought it unlikely.

On 30 November 1855, acceding to a request by Robert Stephenson for advice on the Victoria Bridge across the St Lawrence River, he wrote: 'After much consideration of the "Victoria Bridge" my impression is that a considerable saving could be effected by increasing to a moderate extent the spans and the weight of iron and diminishing the number of piers.'

Progress on the Royal Albert Bridge across the Tamer was slow. In October 1855 the assigned contractor failed. The company took over, and by late autumn 1856 the masonry section of the central pillar had been built to its full height of 12 feet above high water, where a temporary cap was placed.

Unlike Robert Stephenson, Brunel did not trust cast iron and for bridge design favoured wrought iron. The two spans over the Tamar were a combination of arch and suspension bridge, based on Brunel's experience of bridges at Windsor, Newport and Chepstow, the arch being a wrought-iron oval tube with two suspension chains, and the outward thrust of the tube on the abutments being counterbalanced by the inward drag of the chains. Unlike Chepstow, where the tube was almost straight, at Saltash it was curved to a radius equal to the arc formed by the suspension chains and the tube was made oval and not round. The width of the tube equalled that of the suspended platform, so, unlike Chepstow, the chains were vertical. Approximately 40 per cent of the chains used had been made over twenty years previously for the Clifton Suspension Bridge, the Cornwall Railway purchasing these as the Bristol company had run out of funds. The deck carrying the railway was suspended by hangers from the arch and chains.

The first completed span, the western, was floated out in position on 1 September 1857, Brunel having gathered this idea from Evans, contractor for the Conway and Menai bridges which he had watched. At the Menai Bridge, a lack of proper communication with the different capstan gangs controlling the towing and checking cables had endangered the project, so at Saltash Brunel directed the operation from a platform in the centre of the span to improve safety. By signals the span was moved out into position and placed in position between the land and the central pier. Water was then admitted to the pontoons supporting the span to that it was lowered to the piers. The whole operation took two hours and was watched in silence by thousands, Brunel having widely publicised this request.

For launching the span, the tow lines were numbered and signals were made by flag:

Heave in	red
Hold on	white
Pay out	blue
Waved gently means	'gently'
Waved violently means	'quickly'

Robert Stephenson had used hydraulic lifting gear to raise his tubes to the necessary height; this involved building tall towers at extra expense. Brunel economised by pushing the tube up from below using three hydraulic presses at each end. The span was then raised incrementally 3 feet at a time, followed by the masonry of the land pier. Time had to be allowed for the masonry to set. On the central pier, where each tube was supported on two octagonal cast-iron columns, temporary packing was used until the height was sufficient for another cast-iron section to be added.

In 1855 an Act was obtained for making a dock at Brentford on the River Thames. Work began in July 1856 and it opened three years later. The dock sides consisted of a series of horizontal arches.

Previously in 1846 a company had been formed to extend the docks at Briton Ferry, near the mouth of the River Neath, but nothing was done until 1851 when the necessary Act was passed. The works were not finished until 1861. The dock sides were not the usual masonry, but inexpensive slopes of furnace slag obtained from smelting works on the River Neath. Jetties were provided at intervals in the dock for loading and unloading.

The South Wales Railway originally intended to terminate at Fishguard, but to avoid the expensive outlay on a harbour there, it was decided to use the natural harbour of Neyland, Milford

Haven, where there was deep water at all times. In 1857 a timber viaduct was built with a pontoon at its end.

In June 1857, in company with Robert Stephenson, Brunel was granted an honorary degree of Doctor of Civil Law from the University of Oxford in recognition of his continued services to both public and private ends.

On 23 February 1858, Brunel reported regarding the ongoing Saltash Bridge project:

> For some time past the Bridge has been lifted and the piers built up at the rate of six feet a week; and at this rate will be up to the level of the tops of the iron columns, or upwards of two-thirds of its intended height, in the course of a fortnight.
>
> The floor or roadway of the Bridge is now about 54 feet above the ordinary spring tides, the total height to which it is to be raised being 100 feet.

The work on the western span was finished on 19 May 1858, but due to Brunel being fully engaged on the launching of the *Great Eastern* and then suffering a breakdown of health, R. P. Brereton, his chief assistant, supervised the floating of the eastern span on 10 July 1858. This was raised to its full height by the middle of December. There were seventeen approach spans, consisting of simple wrought-iron girders on masonry piers – in fact, all the bridge piers were of masonry except the central one. This comprised four octagonal cast-iron columns built up in sections.

On 23 February 1859, Brunel reported that the works were complete and 'in two or three days the bridge will be tested by the running of heavy trains across it'. It had cost only £225,000, whereas Stephenson's comparable Britannia Bridge over the Menai Strait cost £601,865. The central pier at Saltash, rising from a depth of 80 feet of water, was the deepest yet encountered anywhere in rail engineering.

The original colour of the bridge was an off-white lead paint. This was applied to halt corrosion because the Cornwall Railway could not afford to have it repainted; Brunel thus applied a lead paint layer to prevent rusting.

The bridge was officially opened by Prince Albert on 2 May 1859. Brunel was not present at the ceremony but shortly before his death he made a visit, being drawn across on a specially prepared platform truck by one of Gooch's locomotives. Soon after his death the directors had an inscription in large raised letters placed over the shore archways: 'I.K. BRUNEL, ENGINEER, 1859.'

For the opening ceremony platforms were erected at each end of the viaduct, one on the Devonshire side for the reception of the royal cortége, and one on the Cornwall side for the accommodation of the gentry. In the rear of the platform on the grassy slope of a hill was stationed a battery of artillery, while every surrounding place from which a view of the bridge could be obtained was thronged with spectators. At about 12.15 p.m. the booming of a royal salute from the citadel and from the flag-ship in the harbour announced that the royal train had arrived.

To the cheers of the spectators Prince Albert did not alight but merely stood on the steps of the royal carriage where he was received by dignitaries. After various addresses were given the royal train proceeded slowly across to the Cornish side. As it passed the central pier, the battery stationed on the hill fired a royal salute. After reaching Saltash station the train continued for about a quarter of a mile to give the royal party an opportunity of seeing the bridge from one of the best vantage points. The prince consort alighted and also walked across the bridge to minutely examine the works both above and below the bridge.

The Cornwall Railway was opened formally on 3 May 1859 and to the public the following day. The occasion was marred by tragedy when, two days later, an accident involving the loss of three lives occurred on the line about 3 miles west of the Royal Albert Bridge. An engine and part of a passenger train was precipitated from a viaduct 28 feet into the mud and water below, killing the driver, fireman and guard. After some delay all the passengers were conveyed from their perilous situation, very few receiving serious injury.

Brunel was fond of using timber viaducts which were relatively cheap to build and maintain and attractive in appearance. As the Cornwall Railway crossed many valleys, thirty-four of these structures were required between Plymouth and Truro, their total length being about four miles. Some were entirely of timber, while others had masonry piers built up to 35 feet below rail level, supporting a timber superstructure. The comparatively low cost of timber viaducts enabled them to be, in certain places, most notably on the Cornwall Railway, economically substituted for embankments.

To prevent the timber from rotting it underwent the earlier mentioned process of kyanisation. While this made the timber less inflammable, as an additional precaution a water tank was placed at each end of a timber viaduct. The wood used was yellow pine from Memel in the Baltic and was both of superb quality and cheap. As gradual deterioration was inevitable, he cunningly devised the structures so that defective pieces could be replaced with the minimum of disruption to traffic. The parts were cut to a standard size.

Skilled bridge gangs carefully examined each viaduct quarterly. When repairs or replacements were required the workers would lower themselves from the bridge decking in bowline loops, swinging perhaps 100 feet above the ground. Each gang consisted of fourteen men, a charge man and two look-outs. No failure of a Brunel timber bridge is ever recorded and the only one to ignite was that across the Usk at Newport – and that was caused by a careless workman even before the bridge was completed.

With operations on so vast a scale it was desirable, both in first construction and in subsequent repairs, to have a uniform dimension for the spans, and the subdivision of 66 feet was determined on by Brunel as being suitable for an economic construction. The subdivision of this length was calculated to allow single whole timbers to suffice as direct support of the longitudinal beams.

The line from Truro to Penzance was constructed by the West Cornwall Railway with Brunel as its engineer. It featured nine timber viaducts, and Brunel tried to effect a further economy by using a rail designed by William Barlow of the Midland Railway. An opened-out 'V' in shape, ballast could be packed inside as well as outside the rail and thus obviate the need of sleepers. It had the additional advantage that the gauge could be altered easily and at virtually no cost. Unfortunately Barlow rails proved 'cheap dear' as they were found unsuited to heavy traffic. As the line was connected to existing standard gauge lines, Brunel was happy for it to be built to that gauge, but works were made sufficiently wide for broad gauge to be added later, which was carried out in due course. On 6 November 1866, the first broad gauge goods ran to Penzance, though Through passenger coaches from Paddington did not reach Penzance until 1 March 1867.

On 30 November 1855, prompted by Robert Stephenson, Brunel leant his expertise to the Canadian Grand Trunk Railway regarding the proposal for a box-girder bridge across the St Lawrence River, observing, 'After much consideration of the "Victoria Bridge" my impression is that a considerable saving could be effected by increasing to a moderate extent the spans and the weight of iron and diminishing the number of piers.'

13

IRON COLOSSUS,
SS *GREAT EASTERN*

One might have thought that previous experience would have taught Brunel that he was not really a ship designer, but to him the failures proved an irresistible challenge. He was convinced that he would get it right with another world's largest ship, the SS *Great Eastern*. He had created a partnership with John Scott Russell, an eminent naval architect and shipbuilder. Russell after graduating gained practical engineering skills and then taught science and mathematics in schools before, at the age of twenty-four, he became Professor of Physics at the University of Edinburgh. He experimented with steam engines and boilers and built a not-too-successful steam carriage which ran briefly on the road between Glasgow and Paisley before its boiler exploded.

A canal company asked Russell to investigate steam propulsion on inland waterways. A typical scientist, he realised that hydrodynamic experiments were required in order to measure a hulk's resistance when passing through water. He built a long trough, a foot wide and a foot deep, and filled it with water. He identified four types of waves: a carrier wave made when a quantity of water is added on an existing water and forms a moving wave along an otherwise undisturbed surface; a ripple wave; an oscillatory wave such as made by a stone dropped into water; and a wave with a broken top. He studied the effectiveness of variously shaped hulls passing through the first and last type of wave. Russell found the most efficient shape to be a sharp bow with a concave curve to the widest part of the hull before tapering towards the stern.

Putting the results of his experiments into practice, he was appointed manager of Robert Duncan's shipyard at Greenock. In 1844, Russell was appointed to the roles of editor of a railway magazine and a daily paper; later in 1847 he opened his own shipyard with partners on the Thames, but built vessels to designs required by his clients, rather than to his own ideas.

In due course Russell became secretary of the Royal Society of Arts. One of the new members recruited by him was Brunel, who rose quickly through the ranks and was appointed vice-president of the society in 1848. They had known each other since at least 1839, when Brunel consulted Russell during the construction of the *Great Britain*. Russell was the prime mover in getting the Great Exhibition of 1851 under way.

The lives of the two men had in many ways been similar. Both showed early promise which faded in middle age. Brunel's successful Great Western Railway and *Great Western* steamship had been followed by the blunders of the atmospheric railway; the fact that most other railway companies did not adopt his broad gauge; the grounding and sale of the SS *Great Britain*; and the suspension of the construction of Clifton Suspension Bridge due to lack of investment. Similarly, Russell's shipbuilding yard turned out no startling new designs and his prickly nature got him turned out of the Great Exhibition committee.

The Eastern Steam Navigation Company had been formed in January 1851 to open new passenger mail steamer routes between England, India, China and Australia. Brunel spotted an opportunity in the overseas market because, as the Suez Canal was yet to be dug, steamers for Britain's important empire in India and Australia had to travel via South Africa. Making this long and circuitous route meant time and money was lost with necessary refuelling, especially as coal shipped out from Britain cost four times as much as at home. The answer, thought Brunel, was to build a vessel sufficiently large to carry enough fuel for a return trip without refuelling, using a single load of coal that could be bought at the relatively cheap British price. In a design of this scale, a single propeller would provide about sixty per cent of the drive and paddle wheels 40 per cent. There was a certain amount of risk implicit in this ambitious project, as contemporary ports may not have been able to accept a vessel of this size and neither may there have been a demand for her capacity. As the ship would be going east, the obvious name would be the *Great Eastern*. He was appointed engineer to the company in July 1851.

John Scott Russell was enthusiastic and believed it to be a wonderful opportunity to make his mark as a shipbuilder. On 22 December 1853, Brunel accepted Russell's tender of £272,000 for the hull, £42,000 for the paddle engines and boilers and £60,000 for the screw engines and boilers. Brunel had estimated the cost at £500,000 rather than £374,200 and perhaps should have queried these figures. Brunel himself invested heavily in the *Great Eastern* venture; his plans for building a house at Watcombe consequently had to be delayed.

Brunel's other supporters were Charles Geach, a Birmingham banker and iron manufacturer who had financed Russell's Thames shipyard, and Henry Thomas Hope, a Conservative MP who had taken part in planning the Great Exhibition. As no single dock was large enough in which to build the vessel, it was constructed for a sideways launch.

Brunel typically took little heed of history and wrote to Russell on 13 July 1852: 'The wisest and safest plan in striking out a new path is to go straight in the direction which we believe to be right, disregarding the small impediments which may appear to be in our way, without yielding in the least to any prejudices now existing.' Regarding a disagreement over the shape of the hull, Brunel presumptuously wrote: 'My fresh ideas had the advantage even of your much greater knowledge – hampered by a little preconceived idea.'

The dimension of the vessel far exceeded anything seen before. It was 690 feet in length and thus twice as long as any previous proven design; it weighed 22,500 tons and was thus six times larger than any other ship; and it could carry 4,000 passengers, or 10,000 troops and 10,000 tons of coal and 3,000 tons of cargo. For propulsion she had paddle wheels, a propeller and six masts for carrying canvas. Brunel used both steam and paddle propulsion, as at that time a single screw or paddle would have had insufficient power to propel a hull of that size.

Brunel contributed two important features to the vessel: a double-skinned hull and a cellular construction of the upper decks. He borrowed this idea from the box-girder construction of the Britannia railway bridge across the Menai Strait to Anglesey, designed by Robert Stephenson. Russell's own input was a flat bottom, longitudinal framing, no keel and transverse bulkheads to offer strength. The bow was a straight vertical stem curving to a typical Russell hull. One of the few innovations copied by later shipbuilders was the placing of public rooms and passenger cabins amidships where her pitching was minimalised.

When designing the *Great Eastern* Brunel achieved a remarkable degree of standardisation in plate size and thicknesses, bar sizes and rivet diameters.

Brunel realised that the *Great Eastern's* captain should not just display good seamanship but 'must be devoted exclusively to the general management of the *whole* system under his control and his attention must not be diverted by frivolous pursuits and unimportant occupations'.

On 5 October 1855, Brunel noted down 'Memoranda for my Own Guidance':

1. Must provide a washhouse and large *Icehouse*, Lamp and Candle room, spirit stores, and inflammable stores-room. Of course the best possible place would be aft – and above deck – can this be?
2. It would be a good thing to have a railway let into the deck on each side along which a truck can carry dinner etc etc from the Kitchen to each saloon.

3. To see what room can be gained into the inner side of the Paddle box as a gangway down to the Kitchen: also for the urinals etc from the deck.
4. A poop for a smoking room and bar.
5. Ventilation and drainage.
6. Gas.
7. By constant observation to lay down position and course of slip: and correct compasses.
8. The Captain's Cabin to be before the mainmast, with a bridge close in front of the mast.
9. A principal entrance through one of the loading ports, and a good staircase.
10. Second Tunnel to communicate aft.
11. Make the chimneys oval.
12. To steer by semaphore from forward with a loud bell to call attention. Such steering however is only for emergencies: the ordinary steering will be as usual by binnacle compass.

Some of Brunel's clever ideas were never integrated into the ship's design. One was to have a stream of surface water constantly pumped up through the observer's cabin so that a change in temperature would immediately indicate the presence of icebergs – a much better scheme than the old-fashioned method of an occasional bucketful being hauled on deck when the officer of the watch remembered. Another clever idea took the form of a device to enable a look-out to keep his eyes open in a gale. It consisted of two vertical tin plates placed one behind the other, diverging from the direction of the wind, with a clear, wide passage between the two sets of plates. The wind became separated by the first two inclined plates and the residue that passed on in a straight line was again subdivided so that at the end of the last set of plates there was no rush of air between them and the look-out was in a perfect calm. Glass would not have served the purpose as it would have become obscured by spray.

On 22 July 1852, Joshua Field had warned Brunel: 'No refinement of construction or attempt at economy of fuel will do half so much good as good stoking, keeping the water in the boilers in a proper state and keeping the tubes clean.' Good stokers maintained an even layer of coal in the furnace, not too thick so that it burned sluggishly or so thin that cold air passing through exposed fire bars. Clinker and ash had to be raked out frequently to keep the fire burning at maximum efficiency.

Brunel and Russell, admittedly two men of great stature and will, could not work together successfully; a clash of personalities was somewhat inevitable. In a letter dated 23 November 1853, Brunel wrote to Russell: 'I have slaved to get the specifications ready – don't you delay the contracts for want of drawings – Besides the first attempt is sure not to satisfy me.' This could hardly be described as a courteous letter to his partner and equal.

Remembering the lack of success of the SS *Great Britain,* investors were reluctant to subscribe to the project. In December 1853, Brunel and Charles Geach supplied sufficient funds from their own pockets to commence construction. This began in earnest in the spring of 1854.

By that spring the composition of the company had changed to such an extent that Brunel was now deeply invested in the company and most of the directors were his own nominees. Completion was anticipated to be two years, but then dissension between Brunel and Geach caused problems. The directors tried to skirt the growing enmity by appointing an intermediary, a resident engineer who would report only to the company, but Brunel refused to accept this solution, writing:

The heavy responsibility of having induced more than half of the present Directors of the Company to join, and the equally heavy responsibility towards the holders of nearly half the capital, must ensure on my part an amount of anxious and constant attention to the whole business of the Company which is rarely given by a professional man to any one subject, and, as it seems to me, ought to command a proportionate degree of confidence, or rather command entire confidence, in me ... The fact is that I never embarked in any one thing to which I have so entirely devoted myself, and to which I have devoted so much time, thought and labour, on the success of which I have staked so much reputation, and to which I have so largely committed myself and those who were disposed to place faith in me; nor was I ever engaged in a work which from its nature required for its conduct and success that it should be entrusted so entirely to my individual management and control ... I cannot act under any supervision, or form part of any system which recognizes any other adviser than myself, or any other source of information than mine, on any question connected with the construction or mode of carrying out practically this great project on which I have staked my character; nor could I continue to act if it could be assumed for a moment that the work required to be looked after by a Director, or anybody but myself or those employed directly by me and for me personally for that purpose. If any doubt ever arises on these points I must cease to be responsible and cease to act.

On receiving this letter the directors thought no more of appointing a resident engineer.

Russell made matters worse when it was found that he had been dishonest. Brunel discovered that the quantity of iron used for the *Great Eastern* and that remaining to hand was 900 tons short of what had been supplied. Russell had been diverting it to other vessels under construction in the yard.

On 24 April 1855, Brunel wrote to Russell: 'I begin to be quite alarmed at the state of your contract – four months are gone and I cannot say that even the designs are completed or even sufficiently settled to justify a single bit of work proceeded with – we shall get into trouble.'

By 12 August 1855, matters were even more serious. Only £65,000 of the contract remained to be paid, yet 4,500 tons of ironwork needed erection, so he wrote again to Russell:

I have tried gentle means first, I must now strengthen the dose a little. If you do not see with me the necessity of shaking off suddenly the drowsiness of sleep that is upon us and feel it so strongly that like the sleepy man just overcome with cold you feel that unless done instantly you are lost – In fact unless, as I say, on *Monday next* we are busy as ants at ten different places now untouched I give it up – but you will do it.

Unfortunately Russell and Brunel's clashing personalities and incompatible management styles continued to make working together difficult. Russell was accustomed to delegating, while Brunel wanted to check everything personally – like all perfectionists he was always dissatisfied and kept wanting to make alterations to the design. By this time Brunel was suffering from the early stages of kidney failure and feeling unwell did not improve his temper. Russell was unbusinesslike and submitted an incorrect tonnage for the completed portion of the *Great Eastern*. Brunel, not one to suffer fools gladly, wrote on 2 October 1855:

How the devil can you say that you satisfied yourself on the weight of the ship, when the figures your Clerk gave you are 1,000 tons less than I make it or than you made it a few months ago. For *shame,* if you are satisfied. I am sorry to give you trouble but I think you will thank me for it. I wish you *were* my obedient servant, I should begin by a little flogging.

The *Quarterly Review* ran a feature with a correspondent observing the construction of the *Great Eastern*:

At first a few enormous poles alone cut the sky-line and arrested [our] attention, then, vast plates of iron, that seemed big enough to form shields for the Gods, reared themselves edgeways at great distances apart, and as months elapsed, a wall of metal slowly rose between him and the horizon.

She is destined to carry eight hundred first class, two thousand second class and one thousand two hundred third class passengers, with a ship's company making a total of four thousand and four hundred people ... From side to side of her hull, she measures 83 ft., the width of Pall Mall. She could just steam to Portland Place, with her paddles scraping the houses on each side ... We might also dwell for a moment upon her mighty larder, or draw a comparison between her and the Ark (which by the way had not half her capacity) as she receives on board her flocks and herds, to furnish fresh meat for the passage. But we believe we have said enough to enable those who have not visited the rising edifice to realise the vast extent of the latest experiment in ship-building.

Brunel had requested information he required for launching the ship in August, but Russell was dilatory in replying. On 2 December 1855, still not having been given the information, he wrote again:

I must beg you to let me have with the least possible delay the correct position of the centre of floatation at the 15 ft draft line and let this calculation be made and tested with the greatest care that no possibility of a mistake can arise. I cannot stand any longer the anxiety I have felt ever since we commenced the ship as to her launching and having now calculated myself her weight and centre of gravity at time of launching I must have her centre of flotation and this your people can do better than mine.

That he had lost confidence in Russell is shown by the fact that the same day he wrote to William Jacomb, Brunel's assistant at Millwall:

I have requested Mr Russell to have made a careful re-calculation of the present position of the centre of flotation at the 15 ft draft, but I should like to have this and some other calculations as to flotation made by ourselves also to check this. How are we off for the means of getting those calculations made? If you want further help you must have it ... To satisfy me they must be made by two different sets of people at two different places, say Millwall and Duke Street. You will see at once the importance of all this when you think that *we have not an inch* to spare in the launching when she must be on a perfectly even keel ...

Brunel eventually received the information and losing no time by 19 December 1855 had prepared detailed plans for the launching cradles. Russell then had the temerity to complain that these would cause him unexpected financial cost. Brunel then reported to the directors:

Mr Russell states that the mode of launching *now* proposed will be much more costly than he had contemplated and brings this forward as one of the reasons why his means are insufficient. I cannot say what Mr R's estimate may now be or what he now believes to have been his former expectations, but I can say quite plainly and decidedly that I know nothing of any alteration of plan whatever but that which was always intended ...

I know what is now passing in Mr R's mind and therefore I shall not affect ignorance of if ... He has once or twice thrown out the idea of letting the ship go, leaving it to descend itself and take its chance as in an ordinary launch. I believe the operation simply impracticable, that is to say that a body of that length drawing 15 ft of water could not launch itself broadside by impetus into such a depth of water – it would stick half-way, even if it sent straight and was uninjured, and would probably cost £50,000 to get it out of the mess and I have told him so. But even if such a thing were possible, the proceeding would in my opinion be simple madness unless the vessel were well insured and you wished to get rid of it.

He went on to reveal that there was a deficiency of 1,200 tons of iron, and that Russell was alarmingly building six other ships on the company's land, and concluded:

Mr Russell, I regret to say, no longer appears to attend either to any friendly representations and entreaties or to my more formal demands and my duty to the Company compels me to state that I see no means of my obtaining proper attention to the terms of the contract otherwise than by refusing to recommend the advance of any more money unless and until those terms are complied with.

Brunel suggested that the Eastern Steam Navigation Company should itself take over building the *Great Eastern*:

I have been thinking over the little I *know* and the great deal that I think and can see as clearly as if I knew it, of Russell's affairs, and I believe that the only step that can save him from *immediate* Bankruptcy, and a Bankruptcy which will prevent him ever rising again, would be an arrangement with his creditors under which his liabilities would be ascertained, the position of each creditor defined and the several works now in hand carried on and no new works undertaken or liability incurred. And if proceeding under inspectors, Mr R. would devote his *mind* and attention to his business – I mean his engineering business – each party providing their own materials and the inspectors attending to the expenditure it is just possible that the several works on hand might be completed and the concern left in a state to carry on business. But unless the creditors are immediately called together and some arrangements made, depend upon it that Bankruptcy, and that not a creditable one, is unavoidable to the ruin of the property in the concern, and I very much fear to the ruin of R.'s character. I think it is much to be regretted that R. seems to have no friend about him to give *strong advice*. I have tried it several times and should not be listened to otherwise I would come and give it, but I think I am better away. R. must feel than he had dreadfully misled me more than anybody and would therefore not be disposed to receive my advice as it would be meant.

Russell had incurred a serious financial difficulty. The Eastern Steam Navigation Company had a right of claim on his estate for a breach of contract, but in fact he had no estate as his creditors and Martin's Bank had prior claims. The partly built *Great Eastern* lay on land held by his mortgagees together with the plant required for its completion. After weeks of discussion, Russell's mortgagees and creditors agreed that the Eastern Steam Company had a right of claim and an agreement was made whereby the company could use the land and plant until 12 August 1857. On 26 May 1856, work on the ship, only a quarter of which had been built, was restarted under the direct supervision of Brunel. At this harrowing time, Brunel was supported, encouraged and helped by Robert Stephenson.

Work on Clifton Suspension Bridge stopped in May 1856 due to lack of cash. The two piers were linked by an iron bar 1,000 feet in length and with a diameter of only 1½ inches, from which a basket was suspended. Brunel accompanied by a lad was the first across to ensure it was safe. The basket moved by gravity and ropes. The Bridge Trustees initially charged five shillings per crossing, but later reduced it to a shilling; in twelve months the receipts were £125.

As the *Great Eastern* was not completed by 12 August 1857 terms were negotiated for keeping the yard until 10 October 1857. The ship was not ready by that extended date so the mortgagees took possession of Russell's yard, and workers were denied access to the ship. This meant that Brunel was forced to attempt to launch the vessel on 3 November 1857, the date of the next spring tide. This was against all his principles as there was no time allowed to test all the tackle.

Meanwhile, for Neyland, Milford Haven, Brunel designed a pontoon for landing passengers, cattle and goods from the *Great Eastern* at all states of the tide. 150 feet long, 42 feet wide and 16 feet deep, it was constructed of wrought-iron plates riveted together and stiffened with angle-irons at regular intervals, divided by strong bulkheads with four water-tight compartments. The bridge forming the roadway to the pontoon was also constructed from wrought-iron plates riveted together and boasted a span of

200 feet. One end rested on shore on strong iron hinges built into solid masonry while the other rested on the deck of the pontoon in order to rise and fall with the tide.

As the *Great Eastern* was nearing completion in 1857, Brunel had asked Robert Stephenson and W. S. Lindsay, while they were visiting the ship, 'How will she pay? If she belonged to you, in what trade would you place her?'

Lindsay's reply, although prophetic, displeased Brunel:

Send her to Brighton, dig out a hole in the beach and bed her stern in it, and if well set she would make a substantial *pier* and her deck a splendid promenade; her hold would make magnificent salt water baths and her 'tween decks a grand hotel, with restaurants, smoking and dancing saloons, and I know not what all. For I do not know any other trade, at present, in which she will be likely to pay so well.

Towards the end of October, Brunel stated to the directors:

As regards the period of the launch, I have for some time past calculated upon being ready by the first tides of next month [November]; and, by the unwearied exertions of those on whose assistance I have depended, with the advantage of unusually fine weather, the principal works required are so far advanced that there seems every prospect of success; but a change in the weather is threatening. The time remaining is short, and comparatively small causes may create such delay as to render it more prudent, if not unavoidable, to postpone operations until the following available tide – namely, that of December 2, as no mere desire to launch on the days supposed to have been fixed will induce me to hurry an operation of such importance, or to omit the precaution of a careful and deliberate examination of all the parts of the arrangements after all the principal works of preparation shall have been completed. Should such postponement prove necessary, or be adopted from prudence, everything having been now prepared, the launch would be on the 2nd of December.

> The ship will not be launched in the ordinary sense of the term, but merely lowered or drawn down to low-water mark, to be thence floated off by a slow and laborious operation, requiring two and possibly three tides, and very probably effected partly in the night, and at no one time offering any very interesting spectacle, or even the excitement of risk.

Brunel's many revisions caused expensive and costly delays. Work ceased for nearly four months in 1856, made even worse by the death of the banker Charles Geach. She was ready for launching in the autumn of 1857 but this presented another problem. The *Great Eastern* was too long to enter the Thames end-on and so had been built parallel to the river; it was intended that she would be moved sideways on a patent slipway designed by Brunel. This slipway consisted of two inclined ways from beneath her to a distance of 300 feet down to the low water mark on the bank of the river, at an inclination of 1 in 12. These ways were about 120 feet wide and the set distance between them also 120 feet. The ways were supported on strong piles forced down into a gravel bed. The actual ways were produced of a triple lattice work of timber baulks on a 2-foot-thick foundation of concrete. The track on which the cradles ran were of heavy bridge rail, similar to that used on the GWR. Brunel originally intended hydraulic launching gear to be used, but the directors ruled this out on the grounds of expense so he had to devise another method. No one before had any experience of moving 12,000 tons.

From 28 October 1858, Brunel lived at the shipyard while the launching preparations were made, taking short naps on a sofa in a small office reserved for his use. To prevent too rapid a descent of the hull, a large checking drum was set at the head of each slipway. In the event of the ship being stuck on the slipways, tackles were rigged at bows and stern, chains from these tackles passing around sheaves secured to barges moored in the river and back to steam winches ashore. Additionally, to give a pull amidships, the four 80-ton, manually operated winches which Brunel had previously used at Chepstow and Saltash were mounted on moored barges. As the signalling arrangements for launching the first span of the Royal Albert Bridge had worked so well two months previously, he repeated these. Special instructions were issued to those associated with the operation:

> Provided the mechanical arrangements should prove efficient the success of the operation will depend entirely upon the perfect regularity and absence of all haste or confusion in each stage of the proceeding and in every department, and to attain this nothing is more essential than *perfect silence*. I would earnestly request, therefore, that the most positive orders be given to the men not to speak a word, and that every endeavour should be made to prevent a sound being heard, except the simple orders quietly and deliberately given by those few who will direct.

In the event, there was not the silence that Brunel had requested, as the directors in an effort to gain income had sold over 3,000 tickets for admission to the yard. Brunel had suggested that four policemen should be obtained, thinking that they would only have to contend with trespassers, not paying visitors, but the officers actually present were ignorant of the portions of the yard to be kept clear. Brunel himself had to go and assist in ordering visitors away from the neighbourhood of the launch, and the crowd soon became so great that the men could not see Brunel's signals.

When Brunel was busy with the ship's launch, he was asked to choose from a list of names for the ship. As he had other more important matters to consider, he retorted that it could be called *Tom Thumb* for all he cared. Actually it was called *Leviathan*, but later became *Great Eastern*.

About 12.30 p.m. on 3 November 1858, Brunel gave the signal. The wedges securing the cradles were removed and at the same time the hand and steam winches were set to work. Two hydraulic rams were then used which immediately moved the ship. The bow cradle moved three feet before being stopped, but the men at the aft checking drum, no doubt distracted by the crowd, failed to halt the slide; the vessel stopped by itself. The steam winches and hydraulic rams at both ends of the ships were put into action, but when the bow winch stripped several teeth Brunel ended the operation for that day. Brunel could not be blamed for the delay for it was exactly what he had warned the press might happen.

Brunel was worried that the land below the cradles might settle and thus distort the bottom plates of the ship. Although it was not possible to float the ship until the next high spring tide on 2 December, it was urgent to get her down to the position for floating. In his usual careful manner, Brunel tested a 10-foot portion of the slipways to a static load of 100 tons, which was double the load it would receive. From the result he concluded that if the ship stopped on the slipways, no serious settlement would occur.

The *Illustrated London News* of 7 November 1857 commented:

We have been accused in late years that England, with all her pretensions to the sovereignty of the seas, has, as regarded the shipbuilding both of her navy and her commercial marine, been outstripped by America.

Every now and then appears on our waters a frigate under the flag of the United States, before which our line-of-battle ships are said to hide their diminished heads; and even in the very dandyism of our naval architecture we are said to have been surpassed by a yacht-building of the New World. Whether this is so, or why it is so, it is not necessary here to inquire; but it will suffice to say that on the east bank of the Thames at Blackwell there has arisen a ship such as the world never saw, and which until now has never been conceived by the genius of an engineer, or carried out by the constructive skill of a builder. In looking into some of the statistics connected with the capabilities if this grand tribute to commerce they seem almost to partake of the fabulous. Imagine a floating machine which is calculated to cut through the waves at a speed of eighteen miles an hour; which will accommodate, in all the comforts of a home which are by possibility attainable at sea, some four thousand five hundred persons, and which, if she had been completed, could have conveyed at once ten thousand soldiers to India; whose captain, from his central post of command, will have to use a telescope to see what is going on at the bow and stern; while the old contrivance for issuing orders, the speaking-trumpet, will be altogether out of date, and valueless in his hands; his voice, even with its aid, could hardly be heard halfway to the stern. He will, therefore, have to signal to his officers by semaphore arms by day, or by coloured lamps by night; and he will also have the electric telegraph ramifying to the engine-rooms and to other places which it may be necessary that his instructions should be instantaneously conveyed. Imagine the manufacture of gas on board, and laid on to all parts of the ship, and the carrying the electric light, which will diffuse a perpetual moonlight around the ship.

England has far outstripped the lagging rivalry of the world. There never was, perhaps, so magnificent a realisation of a magnificent idea. We are a colonial nation, and if it were necessary, here are the means of conveying a whole colony of people, with all their means and appliances, at once.

The launching of such an enormous vessel into the river at a point where it is not much wider than her own length across, was a novel experiment, and of course the greatest interest was expressed in the success of an attempt to run a ship sideways into the water until she floated. If such a mass of iron and wood, weighing thousands of tons, could be precipitated into the water with comparative ease and facility, the launch would constitute as remarkable an event as any in the calendar of human invention.

The next attempt to move the *Great Eastern* occurred on 19 November. Two more hydraulic presses were made available. After a few moments strain the chain of the bow-hauling tackle failed, so operations were deferred until 28 November. By the end of the day she had moved only 14 feet, due mainly to weak links in the tackle chains breaking despite theoretically being sufficiently strong.

Brunel realised that it would be best not to rely on the tackles and so devised an alternative method of moving her by simply using hydraulic rams. By 30 November she had moved 34 feet before one of the rams burst putting paid to the idea of a launch on the spring tide of 2 December. On 4 December two more of the press cylinders burst.

It was then found that an increasing amount of pressure was required to start the vessel moving. Robert Stephenson agreed with Brunel that it would be wiser to defer operations until more and stronger hydraulic power was available. While awaiting the new presses which were being manufactured, Brunel received so many suggestions on how *Great Eastern* should be launched that he drafted a standard reply.

The *Mechanic's Magazine*, in its issue of 15 December 1857, was critical:

> Mr Brunel has not been altogether unfamiliar with failures; but no failure of his ever did so much to lower the reputation of English engineers as the launch of the *Leviathan*. Having first, by the construction of that enormous vessel, concentrated the attention of the world upon him, he has now presented to it the greatest and most costly example of professional folly that was ever seen ... it was, in our judgement, an altogether unnecessary display of self-confidence in Mr Brunel to build the ship where she is, particularly as the narrowness of the river and the populousness of its banks rendered a rapid launch extremely dangerous.

In the effort Brunel managed to assemble eighteen presses with a combined thrust of 4,500 tons. A large number of these presses had been lent free of charge. On 4 January 1858, he put them to work, but this time the weather was his enemy – the fog was so thick that his signals could not be seen, thus rendering coordination between gangs impossible. By 10 January she was down the slipways sufficiently to be semi-afloat at high tides. As the water partly supported her weight, progress was much quickened, and on 14 January Brunel stopped operations in case the high tides of 19 January should float her prematurely. She was then pushed until the cradles were 25 feet from the end of the slipways. To prevent the ship floating too soon, Brunel had water pumped into the double hull. Every care was taken: the wedges securing the vessel in the cradles were attached to the ship by chains so that when released they could be hauled up out of harm's way and would not foul anything.

30 January 1858 was the momentous day of the launch. As the weather was so important, Brunel arranged for observers at Plymouth and Liverpool to advise him of conditions by telegram. His son Henry was given leave of absence from Harrow School to watch the operation. Other dignitaries among the onlookers were Prince Albert and the sixteen-year-old Prince of Wales. On 29 January the wind was increasing in strength and reports from Plymouth and Liverpool were discouraging; nevertheless, he decided to go ahead and gave the order to pump out the 2,700 tons of water ballast.

By the early hours of 30 January 1858 a gale was coming from the south-west. Even if she had floated, the power of the tugs would have been insufficient to haul her across the river against the wind. At 3.30 a.m. he ordered the pumps to stop. The water ballast was replaced – a fortunate precaution on Brunel's part, as the following night's tide proved higher than normal and the ballast was just sufficient to hold her. Early on 31 January the wind moved to north-east and Liverpool reported an improvement in weather conditions. He ordered the pumps to restart and, later on, the removal of the wedges. All was ready and he just had to await the tide. At 1.00 p.m. Mary Brunel and Sophia Hawes arrived from Barge House and were shown by Henry Brunel to a good vantage point.

At 1.42 p.m. the great ship was afloat. In the excitement, the stern mooring cradle was released too early and the two front tugs hauled the ship ahead before she had cleared the cradles, causing the port paddle wheel to foul the timbers of the forward cradle. She had to be carefully manoeuvred astern. The starboard paddle wheel fouled a moored barge and Brunel ordered her to be scuttled. The four tugs then drew her across the Thames to her Deptford moorings, the whole operation being a tribute to

Brunel's careful planning for every contingency. At 7.00 p.m. he descended from the ship having been awake for sixty hours.

It was forty-nine years before another ship, the *Lusitania*, would outstrip her in size. The *Great Eastern* was six times the size of the largest line-of-battle ship in service and could accommodate 4,000 passengers or 10,000 troops. Russell had constructed the *Great Eastern* so that no less than 360 10-inch guns could be placed on board if the government should at any time desire to requisition and convert her into a man-of-war, either temporarily or otherwise. She was proof to ordinary round shells. Her bow was sufficiently strong to withstand ramming and her watertight compartments made her difficult to sink.

On 1 February 1858, Robert Stephenson, prevented by illness from being present at the launching, wrote:

My Dear Brunel,
I slept last night like a top, after I had received your message.

I got desperately anxious all day, but my doctor would not permit me to venture as far away as Millwall.

I do, my good friend, most sincerely congratulate you on the arrival of the conclusion of your anxieties.

On 2 February 1858, Brunel wrote to his son Isambard giving details of the launching and ending with a paragraph which shows something of his spiritual side:

Finally, let me impress upon you the advantage of *prayer*. I am not prepared to say that the prayers of individuals can be separately and individually granted, that would seem to be incompatible with the regular movements of the mechanism of the Universe, and it would seem impossible to explain why prayer should not be granted, now refused; but of this I can assure you, that I have ever, in my difficulties, prayed fervently, and that – in the end – my prayers have been, or have *appeared* to me to be, granted, and that I have received great comfort.

14

AN ENGINEER TO THE END
1859

Although she was now afloat, both the company and Brunel's health were ruined by the immense ordeal. He wrote in February:

> For the last 3 or 4 months I have had so much to try my temper that I am proof against anything, and only fear my becoming too slow and apathetic in this state of mind. I would not trouble you with an invalid's journal ... being weak, I am irregularly floored with a concatenation of evils.

With his wife to help his deteriorating health, Brunel left for Vichy in May 1858, and after some time there went on to Switzerland. He was so charmed with his trip up the Rigi that instead of spending just one night he spent a whole week there; Brunel then progressed to Holland. While away ostensibly for pleasure he was, however, busy working on plans for the Eastern Bengal Railway – and no doubt also thinking of the *Great Eastern* on which £732,000 had been spent.

In 1858 work started on Brunel's dock scheme at Briton Ferry. It consisted of an outer tidal basin and an inner harbour, the lock gate between the two forming 56 feet in length and of the buoyant type as used at Bristol and Plymouth. The work was not completed until 1861.

When Brunel returned to England in September 1858 the *Great Eastern* was still at Deptford. Partly because of a trade recession, the company had failed to raise the £172,000 required for completion. It seemed as though she would have to be sold. *The Times*, which had always been hostile to Marc and Brunel, commented with its established line in July 1858:

> We have already had such specimens as the Thames Tunnel and other enterprises which, however they may have redounded to the profit and glory of individuals, have led to the ruin of those who embarked in them on the only principle which should ever be recognized in such cases – namely that of securing an adequate pecuniary return. It may be fine to say that the Thames Tunnel, *Great Eastern* and other analogous constructions excite the wonder of foreigners and should gratify our pride, but in that case we should consider the establishment of a 'Consolidated fund for the Dignity of England'.

On 6 November 1858, Brunel wrote to Thomas Brassey, a director of the new group which hoped to buy the ship:

> It is generally said that the new company is buying this ship under the impression that they are embarking only on a speculation similar to that of buying very cheap a large and costly but ordinary building in which no expense has been spared but which is not quite finished and that all they have to do is to call in a few ordinary tradesmen, upholsterers and painters and finish it off and let it at a large profit …
>
> Now if that is the belief in which the leaders of the new company embark in the matter they will find when too late that a very great mistake has been made.
>
> The mercantile success of the undertaking depends first entirely on the perfect mechanical success of the ship as a machine and though great progress has been made in the construction of the machine the buyers will find that it is not quite such a simple straightforward business to finish it and

make it work efficiently and profitably. Like a half-finished chronometer it may be capable as I believe it is of being made into a perfect machine and then into a profitable one, but it will be much easier to spend a great deal of money and time and make it a total failure. I have no doubt there are many men in England quite as competent as myself to have originally designed the whole, but there are few if any who could now take it up and learn all the difficulties and what may be called the weak points – and make the thing a success.

On 18 November 1858, this new company assumed the name of the Great Ship Company. It was financially floated to purchase the ship and allot shares to the original subscribers in proportion to their holdings. It acquired the hull for £165,000.

Brunel was asked to submit an estimate for fitting out the *Great Eastern*. He produced not only an estimate but detailed specifications, to which he said the contractors should be strictly bound. The pressing need for economy necessitated the abandonment of interesting features such as two small steamers fixed in davits in front of the paddle wheels. These would have acted as tenders at a port which lacked harbour facilities for so large a vessel. Another rejection was a device to provide an artificial horizon and enable astronomical observation to be made no matter what the ship's movement. His plan for special charts on reels, to have been unwound by the ship's patent log as she steamed along, was another feature that had to be summarily abandoned.

Brunel's health failed to improve after this episode, and a consultation with Sir Benjamin Brodie and Dr Richard Bright revealed that he had the disease named after the doctor – latterly known as nephritis. They insisted Brunel winter in a warmer climate, and a few days after the tenders for completing the *Great Eastern* had been invited he left for Egypt, accompanied by his wife Mary and one Dr Parkes, superintendent of the hospital Brunel had designed for Renkioi.

At the beginning of December 1858 they left for Alexandria, collecting Brunel's son Henry in Switzerland and taking a paddle steamer through the Mediterranean. While the rest of the party were being seasick, Brunel was busy measuring the pitching and rolling of the vessel and the velocity of the wind. Christmas Day saw him dining at Cairo with Robert Stephenson. Brunel commenced a journey up the Nile on 30 December, and on 21 January 1859 arrived at Thebes, where he was well enough to ride a donkey. They reached Aswan on 2 February and made preparations for ascending the cataracts. His boat, *Florence*, being of iron could not safely go up the cataracts, so he hired a boat laden with dates, emptied her and fitted up three cabins. Brunel recorded in a letter to his sister on 12 February 1859: 'They really do drag the boats up rushes of water which, until I had seen it, and had then calculated the power required, I should have imprudently have said could not be effected ... These efforts up the different falls had been going on for nearly eight hours.' The Brunels then travelled to Naples and Rome, where he spent Easter before returning to England mid-May.

Meanwhile, the business of the *Great Eastern* was ongoing. On the day he left for Egypt he learned that the Great Ship Company had received two tenders for fitting, one from Wigram & Lucas, who accepted Brunel's specifications, and one from John Scott Russell who did not. On 6 December 1858, Brunel wrote a letter from Lyons urging the board to accept the offer of Wigram & Lucas and not to agree to any contract without the stated specifications. In view of Brunel's prolonged absence and the fact that Russell knew more of the ship than anyone, the contract however was given to him.

Trouble started almost at once. When a complication occurred, Russell claimed it was the company's responsibility. Disputes were so serious that in June 1859 John Hawkshaw, John Fowler and J. R. McClean were called in as arbitrators.

Brunel realised his days were numbered and hoped he could fulfil his ambition of accompanying his great ship on her maiden voyage. Whenever he was able he was on the ship checking the details; and when physical weakness prevented him from going out, he lay on his bed in Duke Street dictating letters concerning such matters as the best coal to use, a reminder to Captain Harrison that he must move to deeper mooring before coaling, that Moulton of Bradford-on-Avon should make the rubber ring joints of the tubular iron masts, and so on. In mid-August, Brunel took the lease of a furnished house at Sydenham – a more convenient base – so that when his health allowed he could visit his ship being fitted out at Deptford.

On 7 September 1859, the *Great Eastern* was to sail to Weymouth. On 5 September, in preparation for this trip, Brunel came on board and at noon suffered a stroke. Paralysed but conscious, he was carried down her side and home to Duke Street. His last instruction before the collapse was that neither he, nor the company, would approve and accept the engines until they had been properly tried at sea. His very last letter on 9 September was a request to the GWR directors that the men from the GWR Swindon works be given time off and a free ticket to see over his ship when it arrived at Weymouth.

To prevent the heat from the funnel making the saloons too hot, and also to double as a feed water heater, a 6-inch-wide jacket around each funnel extended the full height of the saloons. An open-ended standpipe ran to the top of each funnel to enable a head of water to overcome the boiler clack valves and feed the boilers. The standpipes prevented steam pressure building in the heaters. When fitting out, the jackets had been given a hydraulic test to 55 pounds per square inch, so stopcocks had to be fitted temporarily to the standpipe outlets. Unfortunately those over the paddle engine boiler room for which Russell was responsible were not removed. Duncan McFarlane, the engineer in charge of the paddle engine, experienced difficulty in maintaining water level in the boilers, so bypassed the feed water heaters and directed delivery straight to the boilers.

As someone had closed the stopcocks at the top of the heaters, or had never opened them, and as McFarlane had now closed them off at the bottom, they were effectively sealed containers in which pressure was rising. As Hastings came into sight, one of the heaters exploded. Fortunately passengers had left the saloon for the dining room, but five firemen were killed. John Arnott, in the paddle engine room, despatched a man to open the other standpipe cock and thus crucially prevented a second explosion.

The Times reported:

> The forward party of *Great Eastern's* deck appeared to spring like a mine, blowing the funnel up in the air. There was a confused roar amid which came the awful crash of timber and iron mingled together in the frightful uproar. Then all was hidden in a sudden rush of steam.
>
> Blinded and almost stunned by the overwhelming concussion, those on the bridge stood motionless in the white vapour till they were reminded of the necessity of seeking shelter by the shower of wreck – glass, gilt work, saloon ornaments, and pieces of wood which began to fall like rain in all directions.

The *Illustrated London News* of 17 September 1859 offered an in-depth report on the *Great Eastern's* departure from the Nore and the subsequent events:

> An unusual degree of interest is invariably attached to any great undertaking, and, whether it is a failure or a success, we cannot divest ourselves of the most sensitive and personal feeling in the matter. As it is with nations, so is it with individuals; and although this undertaking itself is entirely independent of our control, yet the sympathy which we manifest for its result proves that we are not insensible to the affairs of other people any more than we are to those of our own. This is remarkably instanced in the case of the great ship which has now commenced the active duties for which she was designed and built, and which augers well for the realisation of all those hopes which we entertained in such a project as her creation. The vast improvements

which have been made in naval architecture, as well as in the application of the paddle and the screw, through the agency of steam power, suggested a further and more extensive development in shipbuilding and propulsion which should be found as effectual to the economy of time and distance by sea as an application of the same agency had accomplished by railroads on the land. The size of the vessel was deemed to be the index of this celerity, and the greater the ship the more scope would be allowed for her machinery, and, as a consequence, the increase of bulk would not counterbalance the increase of power which that machinery would apply. The power would naturally be increased to propel the additional weight, but the power itself would be far beyond the proportions necessary only to that end, and must, *per se*, add to the means of effecting a more rapid transition through the waters. If this rule would apply where the larger scope for additional canvas to a sailing-ship enabled mariners to curtail the length of any particular voyage, *a fortiori* must it be demonstrable when the irresistible agency of steam is in greater proportions introduced for the increase in velocity of the means of transit. Many difficulties have been suggested, and some which appeared practically insuperable for the general use of large ships, as instanced in the case of the *Renfrew* and *Columbus,* both of which were wrecked, and the extreme length of each of which was 370 feet, and the width 60 feet. But this was prior to the introduction of steam. When, however, we bear in mind the difference between a sailing-craft and a steamer we can account for the deficiencies in the former, which appear valueless in contrast with the latter. The one is dependent to a great extent on the direction of the wind and the consequent casualties thereof; the other is independent of either and will work her own way against all obstacles. By the agency which baffles the weather and defies the storm. Mr Brunel suggested the plan of the *Great Eastern,* and a company, as we are well aware, was soon formed to carry out the project. That company was, subsequent to the launching of the vessel, dissolved, and a new one established, under whose auspices she has attained her present position, and whose first essay on the sea from the Nore on Thursday, the 8th of September, we briefly chronicle.

All Wednesday night we were disturbed by the continued reverberation of the workman's hammer, but, anticipating as we did such pleasure with the coming morn, we felt no inclination for sleep. The clock had scarcely gone six when the visitors on board were leaving their berths, and coming on deck to see the needful preparations for her independent action when her steam-tugs should be dispensed with, and released from her leading-strings, the long-nursed pet of the ship builders would be left to run alone. The morning looked lowering, but the busy scene in which all on board more or less participated dismissed every other thought than that of the approaching experiment. The pretty little steamer, the *Widgeon*, which had conveyed a portion of her guests from Strood the evening before, lay waiting in attendance as a nurse who was giving up the custody of her child, and who had first taught it the use of its natural power of motion.

Another steamer was at hand to carry despatches to London, and such of the guests as would not proceed beyond 'The Nore'. Some were expressing regret for one cause, and some for another, that they could not accompany her; but all (though that all was but few in number) were equally agreed in the wish to stay, and, as they had witnessed her first movement on the Thames, so did they long to keep with her till she should again drop anchor two hundred miles away.

On the upper deck and at the bow were heard a faint sound of the music of some semi-musical piper bringing out in unmeasured notes the tunes of 'The King of the Cannibal Islands', 'Charlie is my Darling,' 'Away with Melancholy,' &c, &c; and ever and anon a hearty cheer. This took us to see the cause of this early excitement, and there we saw about eighty men working like boys on a 'whirligig' at a country fair at levers placed in the capstan, and around which they were sometimes struggling to make a move, and at others running with great alacrity. This was to draw up the great

anchor; and among the merry group of sailors was many
an amateur enjoying the fun as a stimulant to the relish of
a good breakfast. Ladies were looking on, and applauding
those amusing efforts of gentlemen and workmen. The
casualty of a broken cog only retarded the work for a
few minutes. The piper piped away more merrily than
ever; the gallant captain beat time, and cheered on his
willing helpmates. It was evident the anchor was coming
up fast; and unless some unforeseen accident occurred the
consummation of their hopes was near at hand. Slowly
and surely it began to emerge from the sea. Down went
one man and then another to adjust the ponderous weight
to the ship's side. Like kittens running up and down
some soft-barked tree those men descended and ascended
the deep sides of the fine-beaked vessel until they had
performed their necessary work. At the stern there was
an examination and adjustment of the compass, so that
all should be in readiness to 'be off' at the given signal.
Anxiety and pleasure were depicted on every countenance.
It was the anxiety of hope, the pleasure of having obtained
such a privilege as to be on board the *Great Eastern* and
watch her primary efforts to make way for herself. Glasses
and telescopes were rubbed and cleaned, and each seemed
anxious to see all that could by possibility be seen.

Suddenly the bugle sounded for another act in the
drama of the day. Our friends, who had worked well at
the capstan – Lords Stafford, Mountcharles, Alfred Paget,
Mr Herbert Ingram, M.P., Mr Scott Russell, Mr Campbell,
Mr Cargill, Captain Lay, *cum multis alis* – heard the
trumpet's sound to another duty, and each on board knew
its note was

Knives and forks rattling,

Sweet music for me.

They worked hard and fast; and as an early cup of coffee in
the confusion of the morning was not easily at command,
there was an eagerness to go to breakfast.

Seated in the spacious saloon, and grouped at different tables with our various friends, the interchange of salutations was cordially given and acknowledged. It was like the enjoyment of a déjeuner in some modern hall of vast dimensions; firm and apparently immovable as Buckingham Palace or Windsor Castle. There, however, seemed to be a slight vibration of the smaller candelabra. The air, which blew in welcome breezes through the open windows, could scarcely be sufficient to effect even that gentle swing in which they were signalling some great event. The ship was in motion – her great paddle-wheels were steadily revolving, her screw was working slowly and effectually, we were passing through the waves at a rate of some eleven or twelve knots an hour. The assurance was too exhilarating to allow any more delay at the breakfast-table. All were anxious to be on deck, and see and judge for themselves that the story of their motion which the little chandeliers had told was really true. Then were the glasses and the telescopes drawn out, and every object watched with intense interest as far as sight, unaided or unassisted, could indicate in the direction of yonder Northern Sea ...

We had just concluded dinner ... when there was throughout the whole vessel a sound of most awful import, quickly followed by the fiendish hissing of disported steam ... the lustred candelabra of the saloon grew lustreless. The beauteous mirrors, whose golden frames had previously hidden from our sight the unseemly iron funnel, were shattered to ten thousand fragments, and all the beauty that adorned this portion of the ship was instantly destroyed. The visitors, terrified at the fearful wreck of all that a few minutes before was so perfect, ornamental and apparently secure, were for a moment paralysed at the occurrence of such a shock ... There was a terrified apprehension as to consequences, for none could even look at the deck – strewed as it was by myriads of fragments of broken glass, uplifted skylights, and the prostrate funnel burst asunder ... Captain Harrison, foremost among his crew and passengers, gave the word of command, and every man was at his post, the hose affixed with incredible speed, the donkey-engine at work to supply water enough to have damped a fire that might have gained ascendancy sufficient to destroy a castle ...

So far as the damage done to the interior decorations and fittings of the ship is concerned we are fully aware that money will repair it; but not so those human lives which unhappily have been sacrificed ... None of the passengers could be reckoned among the dead, the injured or the dying. All, with two exceptions, were found safe, and these had but slight scars and bruises. One party of four was sitting in a cabin within five yards of one of the funnels on which was the framework and the plate glass that was smashed to atoms. Some of the glass went like a shower of shot and stone through the apartment where they were sitting, and none were injured ... To the engine-rooms and to the stokeholes was naturally the point of destination for the humane and the benevolent. It was a truly painful sight to see the poor fellows borne along in an almost senseless state of agonising stupor, blackened and disfigured; but it was consolatory to know that the ship contained every possible requisite of medical stores and surgical bandages, from the simple disinfecting fluid to the rarest and most costly medicine. Nor had the directors been unmindful of the requirements of such a vessel going out to sea. Dr Wilson and Mr Slater, two medical gentlemen not only of great skill and experience in their profession, but of the most humane and tender dispositions, with Mr Evans, the assistant-surgeon, formed the staff which was to supervise the health and minister in the casualties of sickness or accident in this isolated home upon the waters. The best mattresses, bolster, pillows and blankets that were in the ship, were soon taken from the luxuriously-furnished cabins of the passengers and made the beds for the poor scalded firemen and stokers ... One more casualty was nearly occurring – not in the ship, but by her – for a ship came dauntlessly and impudently across her path, and a sharp turn of the rudder caused on of the tiller ropes to snap asunder; but the vigilance and promptitude of our pilots saved the lives of the adventurers and their craft, and caused us only temporary inconvenience, for the prudence and forethought of Captain Harrison had guarded against such a probable occurrence, and provided a remedy, in strong iron chains in the event of such a disaster.

When we dropped anchor at Portland the boats and steamers brought their overcrowded freight of happy-looking friends to offer their congratulations … The telegraphic wires were soon at work after some of our party had reached Weymouth … [The *Great Eastern*] has proved herself equal to the highest rate of speed, perfectly tractable and docile, easy even and unshaken in the face of wind and storm, and successful beyond any parallel in the world to ensure her passengers a short or long voyage without the pain, the nausea, and the unpleasantness of that greatest scourge to sea-travelling – sea-sickness.

The *Globe* in its edition of 12 September 1859 revealed that, had not many passengers been on deck admiring the scenery and not in the grand saloon, the death toll would have been far higher. It explained that:

It was an explosion of the "casing," an outer tube which covered the lower part of the funnels or chimneys in the lower part of the vessel, with a supply of cold water in the space between the true chimney shaft and the outer shaft to cut off the heat, and keep down the temperature of the cabins. According to rule, this water should be constantly let off and resupplied; but, by somnolent, the routine was arrested, steam was generated, and hence the explosion. The arrangement had before proved liable to this kind of accident, and Messrs Watt and Bolton had declined to allow the three funnels of their construction to be thus encased.

Passengers made a collection of £22 12s 6d for the families of the dead and injured. The two bodies were taken ashore to await the coroner's inquest and the three surviving sufferers removed to the hospital at Weymouth. This calamity was to be one of the final tragic events of Brunel's life and career.

Brunel died at 10.30 p.m. on 15 September 1859, suffering from Bright's Disease. This kidney disease had led to renal failure; in those days, dialysis or transplant treatment were not available. On 20 September 1859 he was buried in the family tomb at Kensal Park, where his father Marc had contributed to the London Necropolis project.

The private funeral left his residence at Duke Street, Westminster at 9.00 a.m. The cortége consisted of four mourning coaches containing his principal relatives 'and the following gentlemen, distinguished in engineering science: Mr Stephenson, Mr Field, Mr Hawkshaw, Mr Walker, Mr Kendall, Mr Charles Manby, and numerous others'. Twelve carriages belonging to Brunel's friends followed the procession. The houses in the immediate vicinity of his home in Duke Street closed their shops and windows in token of respect.

Along the road to the chapel hundreds of people – including his private and professional friends, neighbours, tradesmen, and the Society of Civil Engineers – assembled. The GWR servants followed the procession to the grave where he was buried with his parents. The sculptor E. W. Wyon took a death mask of Brunel's features in order to prepare a bust.

Perhaps the most moving tribute to Brunel was delivered by Sir Daniel Gooch:

On the 15th September I lost my oldest and best friend … By his death the greatest of England's engineers was lost, the man with the greatest originality of thought and power of execution, bold in his plans but right. The commercial world thought him extravagant; but although he was so, great things are not done by those who sit down and count the cost of every thought and act. He was a true and sincere friend, a man of the highest honour, and his loss was deplored by all who had the pleasure to know him.

The *Bath Chronicle* of 22 September 1859 printed a typical obituary:

We regret to announce the demise of Mr Brunel, the eminent civil engineer, who died on Thursday night, at his residence, Duke Street, Westminster. The lamented gentleman was brought home from the *Great Eastern* steamship, at mid-day, on the 5th inst., in a very alarming condition, having been seized with paralysis, induced, it was believed, by over mental anxiety. Mr Brunel, in spite of the most skilful medical treatment, continued to sink, and at half past 10, on Thursday night, he expired at the comparatively early age of 54 years.

Some local newspapers quoted an article which had appeared in the *Illustrated News of the World* giving a brief biography of Brunel and paying the tribute:

The life of Mr Brunel, no less than that of his distinguished and lamented father, is a standing proof of what may be accomplished in the 19th century by a genius of a high order, seconded by perseverance and industry; he has great claims upon the gratitude of the nation and account of the high character of his professional labours and their essential world-wide usefulness. If, in the words of a French writer, 'the father deservedly gained the celebrity which distinguished him, together with the admiration of all men of learning and of labour, and the affectionate regard of all those who were fortunate enough to know him personally, and to appreciate his simple and noble character,' it is a satisfaction to know that these virtues belong equally to the son, and that wherever the name of Brunel is known, it has secured the esteem and respect of all who have been brought into contact with him.

The *Illustrated London News* of 24 September 1859 recorded:

Isambard Kingdom Brunel, whose death we had the melancholy task of recording in our second edition last week … was on board the Great Eastern (his last important work) on the day before the vessel left the Thames, and remained for several hours to witness the trial of the engines. Symptoms of paralysis showed themselves, and he was hurried home, and laid on his bed from which he never rose again. The news of the explosion on board the great ship reached him on Tuesday last, and from that time he gradually sank until Thursday night, the 15th instant when he expired at 10.30 pm.

The Times of 19 September 1859 carried an obituary notice and a report of the inquest of the firemen. The writer of the obituary gave Brunel 'the conception and design in all its detail of the *Great Eastern*', yet Russell claimed at the inquest that 'except as far as the late Mr Brunel was the originator of the idea, I was the builder and designer of the Great Ship'. The jury at the inquest brought in a verdict of accidental death.

At the time of his death in 1859, Brunel was engaged with designs for the Eastern Bengal Railway. The rails he sketched in 1858 for this line were an interesting new development. Joseph Locke had designed a double-headed rail so than when the upper surface became worn the rails could be taken out and reversed to provide a new running surface. The plan did not work because the lower surface when resting on a chair sustained abrasions and thus offered rough riding. Brunel modified this idea: his double-headed rail did not rest on chairs but was clamped into them, the base not touching the chair and so the under surface remaining undamaged.

At Portland in October 1859 the 7-ton Trotman's anchor on the *Great Eastern* had to be raised by about eighty men, the steam-worked apparatus failing. The screw boilers consumed about 160 tons of coal daily, and the paddle boilers 100 tons. The engines worked perfectly and there were no heated bearings. At 11.00 a.m. the ship's bell rang for divine service and the company on board, numbering about fifty, assembled in the dining saloon where the service was read by the Revd Nicholson, a large shareholder in the company, while the purser, Captain Lay, according to custom, said the responses.

The *Great Eastern* ran from Portland to Holyhead in forty hours at an average speed of almost 13 knots, or 15 mph – and this notwithstanding her bad trim and the fact that for the two nights she was out, due to the engines being stiff and not fully run-in, speed was kept well below what she was capable of. Her normal speed of 15 knots would have enabled her to travel from England to Calcutta via the Cape in thirty-two days, and with no delays needed for coaling. This meant that it would take less time than the overland route and offer greater comfort for travellers.

When leaving Holyhead for Southampton, as the steam gear for raising the anchor was useless, manual labour was applied. The anchor stuck fast; eventually, it snapped in half. Steam had been raised in all of the paddle boilers and four of those for the screw engines, but for approximately an hour the ship was only moved by the screw, as difficulty had been experienced in getting the paddle engines to take their vacuum. The problem continued until the size of the condenser was altered.

The *Great Eastern* suffered one problem after another. The white-metal stern bearing of the screw shaft rapidly wore despite lubrication, while another design error was found in that the lifeboats were set insufficiently high and were consequently damaged in a severe storm. When under steam she was several knots slower than was anticipated because her engines lacked sufficient power; she wagged excessively and, like the *Great Britain*, rolled severely. She crossed the Atlantic several times to New York, but was uneconomic when compared with ships that burned less coal and had a smaller crew.

An Atlantic gale ripped off her paddles wheels and rudder and she ran aground on Long Island, ripping a great hole in her bottom. The fact that she had a double skin saved her. She was usefully employed laying Atlantic cables, as no other vessel had the capacity for the length required to cross the ocean. She was eventually scrapped in 1889.

Brunel had been Vice-President of the Institution of Civil Engineers and of the Society of Arts; a Fellow of the Astronomical, Geological and Geographical Societies and Chevalier of the Legion of Honour.

The Institution of Civil Engineers sponsored the completion of Clifton Suspension Bridge as a fitting memorial in Brunel's honour – it opened 8 December 1864. Engineering friends erected a bronze figure on the Embankment near Temple station, while his family contributed a memorial stained-glass window in the north aisle of Westminster Abbey, paid for by melting down the silver table-centre presented to Brunel by the GWR. His estate was valued at £90,000 – whereas his contemporary Joseph Locke left £350,000 and Robert Stephenson £400,000. Brunel believed that fame and perfection were more important than making money for himself or shareholders.

Brunel had achieved much in his lifetime. He had built more than 1,000 miles of railway, designed over 100 bridges, made the first prefabricated field hospital and revolutionised ship design; Britain was richer for having him as a citizen. His guiding and lasting philosophy was that creation was more important than its commercial consequences.

APPENDIX 1

SIGNIFICANT DATES

9 April 1806 – The birth of Isambard Kingdom Brunel.

2 March 1825 – Work begins on the Thames Tunnel.

3 January 1827 – Brunel appointed resident engineer to the Thames Tunnel.

12 January 1828 – Brunel nearly drowns in Thames Tunnel.

10 June 1830 – Brunel elected fellow of the Royal Society.

31 March 183 – Brunel appointed engineer to the Clifton Bridge.

7 March 1833 – Brunel appointed engineer to the Great Western Railway.

5 July 1836 – Brunel marries Mary Horsley.

27 August 1836 – Foundation stone of Clifton Suspension Bridge laid.

19 July 1837 – SS *Great Western* launched.

31 May 1838 – The GWR opened from Paddington to Maidenhead.

30 June 1841 – First train runs from Paddington to Bristol.

19 July 1843 – SS *Great Britain* launched.

13 September 1847 – First atmospheric train runs between Exeter and Teignmouth.

31 January 1858 – SS *Great Eastern* launched.

2 May 1859 – The Royal Albert Bridge opened.

5 September 1859 – Brunel suffers a stroke.

15 September 1859 – Brunel dies.

20 September 1859 – Brunel is buried in Kensal Green Cemetery.

APPENDIX 2

BRUNEL'S MAIN ENGINEERING PROJECTS

Bridges
Balmoral, Chepstow, Clifton, Hungerford, Royal Albert, Windsor.

Docks
Brentford, Bristol, Briton Ferry, Monkwearmouth, Plymouth.

Hospital
Renkioi.

Railways
Bristol & Exeter; Bristol & Gloucester; Cheltenham & Great Western Union; Cornwall; Great Western; GWR Newbury branch; Indian; Irish; Italian; Oxford, Worcester & Wolverhampton; South Devon; South Wales; Taff Vale; West Cornwall; Wilts, Somerset & Weymouth.

Ships
Great Britain; Great Eastern; Great Western.
The Gaz engine.
Kensington Observatory.
The Thames Tunnel

APPENDIX 3

DIMENSIONS OF SHIPS

	Great Western	*Great Britain*	*Great Eastern*
Length overall	236 ft	322 ft	692 ft
Breadth of hull	35.3 ft	51 ft	82.7 ft
Draught	16.7 ft	18 ft	30 ft
Gross register	1,340 tons	3,720 tons	18,915 tons
Displacement	2,300 tons	3,618 tons	32,000 tons
Boiler pressure	5 p.s.i.	15 p.s.i.	25 p.s.i.
Grate area	202 sq. ft.	360 sq. ft.	2,328 sq. ft.

APPENDIX 4

THE ROOMS IN 17 AND 18 DUKE STREET

John Horsley, Brunel's brother-in-law, wrote that Brunel spent 'the pleasantest of his leisure moments in decorating [the house in Duke Street], and well do I remember our visits in search of rare furniture, china, bronzes, &, with which he filled it, till it became one of the most remarkable and attractive houses in London'.

No. 17
Governess' room
Closet adjoining
Laundry
Lumber room
Cook's room
Mr Isambard's room (Brunel's son)
Landing
Schoolroom
Red room or study
Laundry
Dining room, contents including: carved sideboard, chimney piece with marble figures, Venetian glass chandelier, two crystal mirrors, silk window curtains, gas candelabra, Indian carpet, paintings:
Midsummer Night's Dream: Landseer
Scene from Henry VIII: Leslie
Julliet [*sic*]: Leslie

Masquerade scene from Henry VIII: Leslie
Sauce and her Dog: Callcott
Death of King Lear: Cope
Macbeth: Stanfield
Forest Scene: Lee
Romeo and Juliet: Horsley

No. 18
Right attic
Left attic
Servants end room
Manservant's room
Housemaid's room
Lady's Maid room
Laundry
Mr Henry Brunel's bedroom
Mrs Horsley's room
Mrs Brunel's bedroom
Spare bedroom
Mr Brunel's dressing room
Study, paintings:
Portrait of Brunel
Portrait of Mrs Brunel
Breakfast room, painting:
Portrait of Brunel
Organ room, paintings:
A Calm: J.C. Horsley
Two Gentlemen of Verona, Egg
Drawing room, paintings:
Italian composition: Sir A. Callcott
[sixteen others by Callcott]
[two portraits by Horsley]
Stairs and entrance
Butler's pantry
Larder
Kitchen
Servants' hall

APPENDIX 5

HOUSEHOLD STAFF, 1851 CENSUS

Footman
Two kitchen maids
Three housemaids
Cook
Housekeeper
Governess
Two children's nurses

APPENDIX 6

SELECTION OF LETTERS REVEALING ASPECTS OF BRUNEL'S LIFE

Polite letter to George Stephenson seeking permission from the London & Birmingham Railway for the GWR to use its Euston station.

53 Parliament Street,
15 October 1835

My Dear Stephenson,
I am requested by my Directors to see you upon the subject of the best means of carrying our rails along your line from the point of junction, Kensal, to the depot, Euston. Our rails being placed at a greater width than yours – I believe – I think this may be done without difficulty. You may perhaps differ from me in this opinion. Have you any objection to talking the thing over with me to tell me the difficulties you foresee, if any? I will endeavour to meet them or be prepared to yield to them.

Yours truly,
I.K. Brunel

A month later, he sends a further letter, though in the event, Stephenson turned down the plan:

53 Parliament Street,
16 November 1835

My Dear Stephenson,
In looking over our sketch of an arrangement respecting the depot there appears to me an omission without which the rest would be unintelligible to anyone but ourselves. I mean the allotment of a certain portion of the ground for the depot of the Great Western Railway Company. I propose therefore that we should insert under the head of Euston Square Depot:-

The space now proposed to be purchased (or purchased as the case may be) to be divided longitudinally into two parts, the western part to form the depot of the Great Western Railway Company and intermediate space to form a passage for access common to both depots.

Under the head of Camden Town depot:-

A portion of the space belonging to the Birmingham Company having frontage to the canal to be set apart for the Great Western Railway Company.

As these are consistent with what we arranged and merely explanatory of those arrangements, I think you will see no objection to the insertion of them.

An immediate reply will oblige.

Yours truly,
I.K. Brunel

Brunel was forced to dismiss unsatisfactory employees. Here is one such letter he sent:

18 Duke Street,
4 July 1836

Sir,
Mr Hammond has forwarded to me a copy of his letter to you of the 30th ult and your reply. The latter is characterised by a degree of temper approaching to insolence – peculiarly improper after the moderate and kind letter addressed to you by Mr Hammond.

I shall not require your services after the quarter which ended on 30th ult.

I am sir,
Your obedient servant,
I.K. Brunel

A letter to Mather, Dixon & Co., locomotive builders:

18 Duke Street
August 31, 1836

Gentlemen,
In reply to your enquiries whether I should entertain any strong objection or preference to either of the different modes of constructing a locomotive engine, namely a multiplying motion [gears], or large diameter driving wheels, or a very short stroke, I should certainly prefer either of the first two to the last. As regards the adoption of a multiplying motion, I am by no means satisfied that it may not be very advantageously employed, although quite aware of the objections generally entertained against it, I should be very glad to see the experiment tried and if judiciously constructed and of excellent workmanship, I think a satisfactory result might be fairly anticipated. I consider driving wheels of the diameter you mention as the alternative.

The price you mention of course includes the tender. When you have determined upon the plan of construction and prepared a general drawing, I shall feel obliged by your informing me. I have nothing to add bit that you will have the goodness to consider the order for one engine to be delivered on the First of May 1837 confirmed and I should have been glad to have given the order for two which I supposed you could have done with almost the same speed as one.

I am, Gentlemen,
Your obedient servant,
I.K. Brunel

Brunel gives a free hand to Sharp, Roberts & Co., locomotive builders, but wishes to check the details. Brunel is seen to be in charge.

18 Duke Street,
26 September 1836

Gentlemen,
I have received your drawing of the engine you propose for our railway. With respect to the general arrangement of the engine I should have preferred larger wheels, but having no decided objection to these with their present dimensions, I prefer leaving you entirely free to adopt your own plans in this respect. There are a few points however upon which I should wish to be satisfied and which are not particularly shewn in your drawings. I allude to the steam passages both as regards their form and dimensions and both the supply and waste pipes and I should feel obliged by your sending me the drawing of them as quickly as possible. In the meantime however you can, I presume, proceed with other parts of the engine. As you may consider the order given subject to the period and place of delivery being somewhat more definitely fixed. We shall want the engine at the London end in ten months (the average of the times you mention).

I observe you propose to make only one engine – is this all you can undertake?

I am, gentlemen,
Your obt. Servt.
I.K. Brunel

An example of Brunel's writing to a Guest, Lewis, the GWR's rail manufacturer; clarity of writing is not one of his strong points.

18 Duke Street,
28 November 1836

My Dear Sir,
I enclosed you a template of the rail we now want and which I have reduced to 6 inches as you recommended – accompanying this is a short statement of our requisites as to quality of metal distinguished from quality of the rail when made all things considered – I should say that I do not stipulate for all or by any other words define the quality of the material except that the rails should be of close equal in quality to No 2 iron and in point of workmanship to possess the qualities required in the annexed statement and to obtain which I suppose would require good metal – independent of the stipulation that it should be equal to No 2. I should wish as soon as the rollers are finished to have ten tons by way of experiment.

I am, Dear Sir,
Yours very truly,
I.K. Brunel

In his letter to Thomas Mellings, locomotive superintendent of the Liverpool & Manchester Railway, Brunel shows that his moral views are that he is strongly against head-hunting. It also reveals how keen he was on untried innovations.

18 Duke Street,
17 April 1837

Sir,
I am glad to hear by your letter of the 15th that you have
formed the plan you mentioned of separating the boiler from
the engine; I shall certainly have some engines made on that
plan. With respect to your former letter which I received after
I had written to you enquiring about your patents – an answer
to it might be construed into an inducement to leave your
present situation and I cannot do this – and although you
may think it hard that you should not endeavour to better
condition, yet so long as you are a servant of a Company,
I can have no communication with you which would not be
sanctioned by them.

I hope you won't think that I am acting harshly towards you,
or that I do not appreciate a great deal of useful information
which I have obtained from you, but I must not act so as to
be open to the imputation, however unfounded, of seducing a
person from the employment of the Liverpool & Manchester
Railway Company and cannot therefore give you any advice
how to act – you must form your own opinion and act
upon it.

I am Sir,
Yours very truly,
I.K. Brunel

Brunel thinks in a very detailed manner, leaving nothing to chance
in this letter to the GWR directors.

18 Duke Street,
Westminster.
1 June 1837.

Gentlemen.

Having considered with Mr Saunders the number of carriages likely to be required during the first six months of working of the railway when opened to Maidenhead, I think it is necessary that immediate steps should be taken to ensure the completion of 55 coaches of all classes and I should propose to divide them in the following proportions and different classes.

It is assumed that there may frequently be 3,500 passengers or journeys to perform in the day and that occasionally there may be 6,000 – and that these may consist of 4 classes and in the proportion:

1st – 1,000
2nd – 1,500
3rd – 1,500
4th – 2,000

The first I suppose to be persons travelling for pleasure or desirous of obtaining any comfort and willing to pay an increased charge and the carriages for their accommodation would be superior to that are usually termed mail coaches first class and might be called private trains or some other distinguishing name – the other three classes being called first, second and third as on the other railways. The number of each of these coaches to accommodate the passengers as classed above would be:

Private 10
1st Class 15
2nd ditto 15
3rd ditto 15
Total 55

And probably about 15 to 20 platforms [ie flat trucks] might be for private carriages and even stage coaches I think may very likely use them. For 1st and 2nd class drawings are in progress and I am prepared if it be considered desirable to advertise at once for tenders for their construction and in order to facilitate this, also to prevent the risk of inconveniences from the introduction at any time of too many new arrangements, I have made them nearly similar to the Birmingham carriages except their being more roomy and comfortable and having large wheels, though not too large as I should ultimately recommend. In the details of the buffers and the connections of the carriages I have introduced some alterations which I consider very material improvements.

With respect to the 3rd class and platform luggage wagons, I am not quite prepared to propose any immediate steps.

I am, Gentlemen,
Your obedient servant,
I.K. Brunel

A letter to his directors revealing his supreme confidence in what proved to be a poor design of track. (As the track was frozen to the ground, he had not made a fair test.)

18 Duke Street,
Westminster.
22 January 1838

Gentlemen,
I am happy to be able to inform you that the trials which have been made upon a portion of permanent rails completed at Drayton have been such as to furnish a decided practical result in the experiment of the application of continuous longitudinal bearings of timber and as a result from which safe conclusions may be drawn as to the advantages of the use of timber compared with the system of separate stone blocks or transverse sleepers.

The peculiarity of the plan which has been adapted consists principally in two points – first in the use of a light flat rail secured to timber and supported over its entire surface instead of a deep, heavy, rail supported only at intervals and depending on its own rigidity – secondly in the timbers which form the support of this rail being secured and held down to the ground so that the hardness and degree of resistance of the surface upon which the timbers rest may be increased by ramming to an almost unlimited extent.

The first, namely the simple application of rails upon longitudinal timbers is not new indeed as mentioned in a former report I believe it is the oldest form of railway in England – but when lately revived and tried upon several different railways it has I think failed & I believe very much from the want of some such means as that which I have adopted for obtaining a solid and equal resistance under every part of the timber and a constant close contact between the timber and the ground as I believe this to be entirely new and to merit to be well understood.

The experiments have been made under several disadvantages and I am glad that it has been so as we are more likely to perceive at once and to remedy any defects which might otherwise have laid concealed for a time. The packing upon which it is evident everything depends was effected during a long and continued wet and while no drainage at all existed, once or twice the timbers and packing were left completely exposed – the severe frost which immediately followed converted the wet sand into a mass of stone. It is still so hard within as to resist the pickaxe and has been with difficulty broken through at some points with a smith's cold chisel and hammer. Under these circumstances with engines weighing between 14 and 15 tons and from want of adjustment with more than half of this occasionally thrown upon one pair of wheels constantly running over the rails.

The timbers have stood most satisfactorily – I think generally there is less motion than I have seen in the best laid blocks – and I am convinced that with good packing, no perceptible yielding would occur indeed such is now the case at several points where the packing is good. upon the whole the result is such that it may now be safely asserted that the objections which have been urged against the use of a continuous support of timber do not exist, certainly in this mode of construction and that no new or unforeseen difficulties exist.

Letter to David McIntosh, contractor, complaining of inaccurate work, but offering to do all he could to help.

18 Duke Street,
2 March 1838

Dear Sir,
Since writing the enclosure I have receive the confirmation of
a report which had reached me of one of your bridges (Wood
East) being nearly 18 inches out of line with three other bridges
close by. There is no alternative: the error is far too great to be
remedied by anything short of rebuilding and not an hour must
be lost if every expedition be used. I will assist by allowing the
wing walls to remain and by preparing a design to enable you
to do this – but this is entirely conditional upon every means
being adopted to expedite the work. On one side it will be merely
adding brickwork which can be done at once.

Yours very truly,
I.K. Brunel

Brunel expresses great sympathy towards Michael Lane, his
resident engineer on the Monkwearmouth Dock project when he
became injured.

18 Duke Street,
29 March 1838

My Dear Lane,
The account of your accident distresses me very much. I must
insist upon your having the best advice that can be procured
without regard to expense. I hardly know who to wish to, to
request them to see that you have everything you want. Do not
trouble yourself about the works. I will think of somebody
between this and tomorrow.

Yours very truly,
I.K. Brunel

Shortly after writing the above letter, Brunel fell off a ladder in the SS *Great Western's* engine room and injured himself. He wrote to Lane:

18 Duke Street,
13 April 1838

My Dear Lane,
In answer to your enquiries I am getting on very well – though very sore and weak from bruises. I have not broken a bone about me and but for excessive weakness, this is the first attempt at writing I being very black and blue and very sore. I should be quite well I hope in a week to be able to attend business again. I wish I could hope you would get on as well.

I was brought home on Saturday.

Yours very truly,
I.K. Brunel

Brunel's enjoyment of a fight and his supreme self-confidence is shown in this letter to his friend Thomas Guppy.

12 July 1838

My Dear Guppy,
A splendid storm is brewing and although we have no umbrella or shelter and must weather it out, one is curious to know beforehand whether it will be snow, hail, or all three – and thunder and lighting to boot – am certain that it is going on to do neither more or less than to condemn all my plans adopted in the GWR and to dispense with the necessity of giving me any further trouble.

I need hardly say that Liverpool is the gun whence all this shot principally comes – our friend Parsons, Sharp and some other of their class are most active. Now I want to learn all I can of what they have to say – then, as to the engineering opinions – will you pick up what you can of the arguments made and opinions given – and names as much as possible? I understand this is no joke, but a serious attack our Directors certainly look upon it much nay almost as 'une affaire finie'.

I am by no means disposed to treat it lightly, though a good attack always warms my blood and raises my spirits.

Yours sincerely,
I.K. Brunel

An engineer, Nicholas Wood, dared to criticise Brunel's track and received this reply:

18 Duke Street,
27 December 1838

Dear Sir,
I regret very much the determination you have come to, it would have been much more agreeable to me that you should have witnessed the experiments I have made and have satisfied yourself of the results and if you had seen occasion to have done so, as I think you must to have qualified the expression of opinions formed upon very different results – rather than that I should be obliged to show that the conclusions to which you came were erroneous in so far as they were founded upon insufficient data.

I am suffering from inflammation in the eyes which prevents me writing to you myself.

I am, my dear sir,
Yours very truly
I.K. Brunel

A reply sent to William Venables, passenger in a train involved in an accident.

18 Duke Street,
February 7 1839

Dear Sir,
I am very much obliged to you for your note. The faults of which you speak of are unfortunately my department and although I have not exactly the power of entirely correcting them, it is my business to render them as few as possible and by using every means reduce the chance of such accidents.

You have been most unfortunate in being twice inconvenienced by them, the more so as I believe you have been tolerably free of such casualties.

The policeman whose business it was to watch the switch in the case of the last accident has been dismissed.

I am, Dear Sir,
Your most obedient servant,
I.K. Brunel

Brunel watched very carefully that the terms of the contract were kept and that only perfect material delivered. He wrote to the GWR directors:

18 Duke Street,
25th Feby. 1839

Gentlemen,
A considerable quantity of rails have now arrived at the Company's wharf at Bull's Bridge and I am sorry to say that a very large proportion – upwards of one half – require straightening or are otherwise defective. A cargo lately arrived from Messrs Harford's appears to have exactly the same defects and in about the same proportion as those from Messrs Guest.

These defective rails are laid aside and unless I receive written directions from you to adopt means to straighten them at your charge, they cannot be included in the certified payment.

I much fear that there will be considerable difficulty in straightening them cold and I hope directions have been sent to the works to use much greater care in the original straightening.

I can carefully assert, from having myself superintended the making of several rails to ascertain the fact that a very little attention to the first operation of straightening while the rail is hot ensures without any expense a perfection which can never be attained afterwards and which it has been expressly stipulated should be attained.

I am, Gentlemen,
Your obt. Servt.,
I.K. Brunel

When his assistant William Glennie was injured in Box Tunnel, Brunel sent this letter to his wife:

June 14th 1839

My Dear Mrs Glennie,
I am much obliged to you for your letter and am delighted to hear that Glennie is improving. I had expected to see you today, but have been prevented from leaving Town by an accident last evening from the effects of which I feel rather stiff. Will you have the kindness to let me hear again on Monday morning.

Yours very truly,
I.K. Brunel

Letter to S. J. Tucker, a young man unable to take criticism and so resigned, but Brunel believed he had a good future and was mistaken to throw away his opportunities.

Bath
22 June 1841

My Dear Sir,

I have been too busy to reply to your letter expressing a wish to leave. A young man in your position would have shewn more wisdom and modesty if he had waited till I had intimated to him that his services were no longer required, instead of assuming, without any real ground for such an assumption, that he was no longer of any use and if I thought it was only foolish or a bit of temper, I should pass it over and tell you to still do your work and not be surprised to find that your particular work was not the most important that I have and consequently that it might without damage to the concern be somewhat neglected while I have very pressing business to attend elsewhere, but I see in this a strong proof that you have not and cannot without infliction, cure yourself of that which I have seen to be the ruin of most young men who, notwithstanding good abilities and good qualities have not succeeded.

You cannot even see yourself in fault and we are constantly forcing yourself into a little higher position than the one you are placed in. I am not a habitual fault finder, on the contrary I fear I am to blame for some of the failures of my Assistants because I dislike complaining. I have not very often been finding fault with your engineering proceedings – it would be much easier with so many young men hanging about me to dismiss than attempt to care.

The probability is that I am in the right when I do complain and you would do much better to look at yourself and try to discern and cure the faults I complain of, than to take the huff like a child and cast away your present means of support and for a good introduction to future work as you would a top because you could not spin it as well. I shall do you a kindness and relieve myself of much trouble which I have taken with you by letting you feel the consequences of your folly – you will remain to the end of the quarter as you wish and I may want you

for a month more. I have appointed Ward to succeed you
at the Engine House.

Yours truly,
I.K. Brunel

Brunel was kind to good employees. He supported Hurst's decision
not to drive a train with an inexperienced guard. Had Hurst done
so it would have been unwise as the gradients were severe. The
only brake power was on the engine and brake van, so for safety
it was essential to have an experienced guard to know when to
brake for stations or gradients. On Hurst's behalf he wrote to the
directors:

18 Duke Street,
18 June 1842

Gentlemen,
I have the satisfaction of submitting to you the names of sixteen
enginemen who have conducted themselves well in every respect
during the last 12 months and upon whom no fines have been
inflicted by you during the whole period, indeed, with one
exception, there has not even been a complaint and in this case
the man was reprimanded.

In addition to the sixteen to whom according to the Rule
Book laid down by you the premiums of £10 became due,
I have strongly to urge upon your favourable consideration the
case of James Hurst.

This man is an excellent servant to the company, useful and
trustworthy in a very high degree, but somewhat tough in his
character. He holds a responsible situation as the head Engineman
in the Cheltenham branch and as such has conducted himself
well. Last year he earned and received the premium of £10.
He had almost done so this year when on 29th December he was
fined 10/- for refusing to take the train from Swindon if Burton,
the sub-Inspector, was sent in charge instead of the regular guard.
I really believe the man considered Burton inefficient & I think

if under the peculiar circumstances of the case he were strongly admonished but the premium allowed, that good servant would be encouraged and no bad precedent established as there are few men who have the same claims as Jas. Hurst.

I am, Gentlemen,
Your obt. Servt.
I.K. Brunel

A letter, probably to Charles Saunders, secretary to the GWR, sent at the beginning of the Railway Mania and offering advice on the proposed Exeter & Exmouth Railway.

October 29 1845

My Dear Sir,
I returned only this afternoon and am endeavouring but very ineffectually as yet to comprehend the present extraordinary state of railway matters – when everybody around seems made – stark, staring, wildly mad. The only course for a sane man is to get out of the way and keep quiet and as a general rule I think it may be safely followed. Whether the Topsham & Exeter concern is to cause an exception is most difficult to say. I am half inclined to think it ought to be – that is to say that if you can unite the really respectable & powerfull [*sic*] interests of Exeter & Exmouth and the landowners & bring them to Countess Weir (or rather as much below as possible) it will no doubt be a great protection to all our interests & a good public work, the B&E are particularly interested in it. The value of the Ferry is not of course worth a moment's consideration. As to Starcross I apprehend the question would always and will still depend upon the merits of the site as regards deep water etc, etc.

There is one thing that strikes me however, the names on the prospectus don't seem to imply any great local strength & we should be sure that we acquire strength and a good deal to compensate for the risks in meddling with such a project.

I am, my dear Sir,
Yours very sincerely,
I.K. Brunel

Brunel pays tribute to his assistant at the Royal Albert Bridge,
Robert Brereton:

Birmingham
30 August 1854

My Dear Saunders,
Brereton, Bertram and I have been at work incessantly, this
bridge affair having involved a serious amount of labour, each
bridge requiring examining and designing and putting out all
the sizes of plates that will be required. Bagnall has undertaken
to roll the iron and I have just finished a preliminary list of
work. It will take them nearly 3 weeks to complete the iron,
working entirely for us and we cannot put piling without it,
and less than 3 weeks to complete the last girder after the last
plate is delivered, so that six weeks is the minimum, but I think
it may, if we are lucky, be done in that time, if organising staff
and putting entirely in charge. Brereton will take it in hand
and indeed without him it would be impossible. I had no idea
myself and you can have no idea of the amount of work that it
has required. We must get all the iron and then get makers to
make the new work. I shall try and put some at Swindon and
at Gloucester and at Chepstow, for there seems little chance of
getting much done here or of getting some workmen, however
we can only do our best.

Yours truly,
I.K. Brunel

BIBLIOGRAPHY

Awdry, C., *Brunel's Broad Gauge Railway* (Sparkford: Oxford Publishing Co; 1992)

Beckett, D., *Brunel's Britain* (Newton Abbott: David & Charles; 1980)

Binding, J., *Brunel's Bristol Temple Meads* (Hersham: OPC; 2001)

Brindle, S. *Brunel* (London: Whitefield & Nicholson; 2005)

Brunel, I., *The Life of Isambard Kingdom Brunel Civil Engineer* (London: Longmans, Green; 1870)

Bryan, T., *Brunel The Great Engineer* (Shepperton; Ian Allan; 1999)

Buchanan, R.A., *Brunel The Life and Times of Isambard Kingdom Brunel* (London: Hambledon and London; 2002)

Buchanan, R.A. & Williams, M., *Brunel's Bristol* (Bristol: Redcliffe; 1982)

Buck, A., *The Little Giant* (Newton Abbot: David & Charles; 1986)

Clements, P., *Marc Isambard Brunel* (London: Longmans; 1970)

Falconer, J. *What's Left of Brunel* (Shepperton: Dial House; 1995)

Fox, S., *The Ocean Railway* (London: HarperCollins; 2003)

Gardner, J., *Brunel's Didcot* (Cheltenham: Runpast; 1996)

Gillings, A., *Brunel*, (London: Haus; 2006)

Jones, R., *Isambard Kingdom Brunel* (Wharncliffe Transport: Barnsley; 2011)

Lewis, B., *Brunel's Timber Bridges and Viaducts* (Hersham: Ian Allan; 2007)

Mitchell, V., *Brunel A Railtour of his Achievements* (Midhurst: Middleton; 2006)

Noble, Lady Celia Brunel, *The Brunels Father and Son* (London: Cobden-Sanderson; 1938)

Pudney, J., *Brunel and his World* (London: Thames & Hudson; 1974)

Pugsley, Sir Alfred, (editor), *The Works of Isambard Kingdom Brunel* (London/ Bristol: Institution of Civil Engineers/University of Bristol; 1976)

Rolt L.T.C., *Isambard Kingdom Brunel* (Harmondsworth: Penguin; 1957)

Rolt L.T.C., *The Centenary of the Royal Albert Bridge* (page 307 ff, May 1959 *Railway Magazine*)

Vaughan, A., *Isambard Kingdom Brunel Engineering Knight-Errant* (London: John Murray; 1991)

Vaughan, A., *The Intemperate Engineer* (Hersham: Ian Allan; 2010)

The Brunel Collection of the University of Bristol holds the largest amount of material on I.K. Brunel and his family. The present author is greatly indebted to the University for providing many of the letters.

INDEX